Summary

Given a familiar object extracted from its surroundings, we humans have little difficulty in recognizing it irrespective of its size, position and orientation in our field of view. Changes in lighting and the effects of perspective also pose no problems. How do we achieve this, and more importantly, how can we get a computer to do this? One very promising approach is to find mathematical functions of an object's image, or of an object's 3D description, that are invariant to the transformations caused by the object's motion. This book is devoted to the theory and practice of such invariant image features, so-called *image invariants,* for planar objects.

Following the introduction in chapter 1, the book discusses features that are invariant to image translations, rotations, to changes in size and in contrast, with particular attention being paid to the effect of using discrete images rather than continuous ones. The next chapter presents a tutorial introduction to the theory of algebraic invariants which lies at the heart of two important types of invariant features: moment invariants for affine transformations, and projective invariants for perspective transformations. Chapter 4 is devoted entirely to features invariant to affine transformations: the theory behind moment-based invariants, Fourier descriptors and differential techniques is presented, along with a novel technique based on correlations, and results of experiments on the stability of coarsely sampled images are discussed. Chapter 5 goes one step further and covers features invariant to perspective transformations, summarizing work on both differential and global invariants. The penultimate chapter, chapter 6, shows how invariant features can be used to recognize objects that have been partially occluded. A thorough treatment of the 'geometric hashing' method is given, followed by some novel methods of 'back-projection' which allow one to verify whether the hypothesized object really is in the image. Many authors claim that moment invariants cannot be used under partial occlusion; this is not so, and a number of schemes for their use are presented. Not only can they be used, but they have some significant advantages over other invariant features, a fact that is backed up by experiments. The final chapter contains a summary and conclusions.

Acknowledgements

I would like to thank Anil Kokoram, Michael Lai and numerous others for ensuring the smooth running of the Novell network in the lab, Mark Wright for helping me to acquire real images, Nick Kingsbury for some suggestions on how to improve the presentation of the material and Roberto Cipolla for proof reading so swiftly and accurately. I would also like to thank Peter Rayner and the United Kingdom Science and Engineering Research Council for giving me the opportunity to carry out the research contained within.

Contents

Chapter 1

Introduction

1.1 Scope

Research in computer vision is aimed at enabling computers to recognize objects without human intervention. Applications are numerous, and include automated inspection of parts in factories, detection of fires at high-risk sites and robot vision, especially for autonomous robots. For the sake of convenience, the task is usually broken up into two stages, 'low-level' vision and 'high-level' vision. Low-level vision involves extracting significant features from the image, such as the outline of an object or regions with the same texture, and often involves segmenting the image into separate 'objects'. The task of high-level vision is then to recognize these objects.

The following is concerned with high-level vision, in particular with finding properties of an image which are invariant to transformations of the image caused by moving an object so as to change its perceived position and orientation, and in some cases its brightness. The idea of invariance arises from our own ability to recognize objects irrespective of such movement — if we look at a book from a number of different orientations, we have no difficulty in recognizing it as a book each time: we can say that a book has properties which are invariant to its size, position and orientation. Finding mathematical functions of an image that are invariant to the above transformations would thus provide us with a technique for recognizing objects using computers, as well as providing us with a possible model for part of human vision. If we wish to recognize an object using a computer, and we assume the computer has stored the models (or example views) of the objects to be recognized in its memory, and the object to be recognized corresponds to one of these models, the straightforward approach searches sequentially through the computer's memory, trying out each model and seeing whether it can be positioned in such a way as to produce an image that matches that of the object to be recognized, until a good match is found. Clearly, this is computationally intensive; ideally, we would like to be able to extract the correct model directly from the information contained in the image — this is precisely what the so-called *image invariants* allow us to do.

Images are the projection of the three-dimensional world onto a two-dimensional (planar) surface, be it the retina or an array of sensors in a video (CCD) camera. Burns *et al.* [1] have proved that one cannot compute an invariant function of the image coordinates of a set of general points in three dimensions (3D) from a single view; one requires at least two views. If one restricts oneself to planar or near-planar objects however, one can obtain a large number of invariants based on a single view, as we

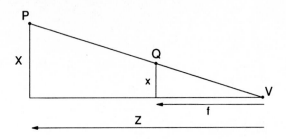

Figure 1.1: The perspective camera

Consider the X-component of a point P in 3D. V is the point of projection, the image plane is defined by Z = f. Q is the image point corresponding to P. x = fX/Z.

will see in the following chapters. Although the theory only applies to flat objects, the resulting features have been used to recognize non-planar objects, such as aircraft, successfully (see references [2, 3, 4, 5]) and will also work well for near-planar objects such as machine parts or the tools used in the experiments described in chapter 6. If one uses two views from an uncalibrated stereo camera, a number of invariants for non-planar points and curves exist — see references [6, 7, 8, 9] and in particular the excellent collection of articles edited by Mundy & Zisserman [10].

1.2   Viewing transformations

The motion of a solid object in 3D is governed by six parameters, three for translations and three for rotations. This section shows how image points are affected by these six parameters, that different views of coplanar object points are related by planar projective transformations and that, if the distance of the coplanar object points from the camera (their *depth*) does not vary much compared with their average depth, the planar projective transformation can be approximated by the affine transformation.

A very good approximation to image formation in a real camera is given by the *perspective camera* model, in which points are projected from 3D onto the image plane so that all the rays joining object and corresponding image points pass through a single point, called the point of projection. If we choose our 3D coordinates so that the origin coincides with the point of projection, the Z-axis is perpendicular to the image plane and points away from the camera, and the image plane is defined by $Z = f$ (see figure 1.1), then the image coordinates (x, y) of a 3D point (X, Y, Z) are given by

$$x = f\frac{X}{Z}; \qquad y = f\frac{Y}{Z}. \tag{1.1}$$

Let $\mathbf{p}^T = [X \ Y \ Z]$ and $\mathbf{p}'^T = [X' \ Y' \ Z']$ be a rotated and translated version, then

$$\mathbf{p}' = \mathbf{R}\mathbf{p} + \mathbf{t},$$

where $\mathbf{R}$ is a 3×3 rotation matrix and $\mathbf{t}$ is the translation vector. This can be written in full as

$$\begin{bmatrix} X' \\ Y' \\ Z' \end{bmatrix} = \begin{bmatrix} r_{11} & r_{12} & r_{13} \\ r_{21} & r_{22} & r_{23} \\ r_{31} & r_{32} & r_{33} \end{bmatrix} \begin{bmatrix} X \\ Y \\ Z \end{bmatrix} + \begin{bmatrix} t_1 \\ t_2 \\ t_3 \end{bmatrix}. \tag{1.2}$$

Combining this with (1.1) allows one to write the image coordinates as a function of the original 3D coordinates and the motion parameters.

If we restrict ourselves to planar objects, Z is related to X and Y: in general, object points lie in a plane $Z = aX + bY + c$ for some constants a, b and c. Putting this into (1.2) gives

$$\begin{aligned} X' &= (r_{11} + ar_{13})X + (r_{12} + br_{13})Y + (cr_{13} + t_1) \\ &= a_{11}X + a_{12}Y + a_{13}, \end{aligned}$$

where $a_{11} = r_{11} + ar_{13}$ etc. Doing the same for Y' and Z' allows us to write (1.2) as

$$\begin{bmatrix} X' \\ Y' \\ Z' \end{bmatrix} = \begin{bmatrix} a_{11} & a_{12} & a_{13} \\ a_{21} & a_{22} & a_{23} \\ a_{31} & a_{32} & a_{33} \end{bmatrix} \begin{bmatrix} X \\ Y \\ 1 \end{bmatrix} \quad \text{or} \quad \mathbf{p}' = \mathbf{A}\mathbf{q}. \tag{1.3}$$

If we set $f = 1$, the image coordinates of the transformed point are given by $x' = X'/Z'$, $y' = Y'/Z'$, from which we see that arbitrarily scaling $\mathbf{A}$ does not affect the image coordinates, which in turn allows one to set $a_{33} = 1$.

The above uses the actual 3D coordinates (X, Y, Z) and (X', Y', Z') of a point and its transformed version; to find invariant functions we must assume no knowledge of 3D coordinates, so we must find the form of the transformation linking image points in one view to those in another. In fact, they are related by a *planar projection*, which is easily proved using *homogeneous coordinates*. Let the image coordinates (x', y') be represented in homogeneous coordinates by (h'_X, h'_Y, h'_Z), with $x' = h'_X/h'_Z$ and $y' = h'_Y/h'_Z$ (we are *not* assuming that $h'_X = X'$, $h'_Y = Y'$ or $h'_Z = Z'$). Furthermore, let $\mathbf{h}'^T = [h'_X \quad h'_Y \quad h'_Z]$. It is clear that $\mathbf{h}' = \alpha\mathbf{p}'$ for some non-zero scalar α, from which we see that $\mathbf{h}' = \mathbf{A}_1\mathbf{q}$, $\mathbf{A}_1 = \alpha\mathbf{A}$. Let $\hat{\mathbf{h}}$ represent a second view of the planar points, then $\hat{\mathbf{h}} = \mathbf{A}_2\mathbf{q}$. If we define $\mathbf{B} = \mathbf{A}_2\mathbf{A}_1^{-1}$ we see that $\hat{\mathbf{h}} = \mathbf{B}\mathbf{h}'$:

$$\begin{bmatrix} \hat{h}_X \\ \hat{h}_Y \\ \hat{h}_Z \end{bmatrix} = \begin{bmatrix} b_{11} & b_{12} & b_{13} \\ b_{21} & b_{22} & b_{23} \\ b_{31} & b_{32} & b_{33} \end{bmatrix} \begin{bmatrix} h'_X \\ h'_Y \\ h'_Z \end{bmatrix},$$

which defines a planar projection (see figure 1.2). As before, we can set $b_{33} = 1$. Since there are only six degrees of freedom, but planar projection has eight free parameters, we see that perspective forms a subset of planar projection. Since the choice of h'_Z is at our disposal, we can set $h'_Z = 1$, which means that the image coordinates $(\hat{x}, \hat{y})$ and (x, y) are related to one another by:

$$\hat{x} = \frac{b_{11}x + b_{12}y + b_{13}}{b_{31}x + b_{32}y + 1}, \qquad \hat{y} = \frac{b_{21}x + b_{22}y + b_{23}}{b_{31}x + b_{32}y + 1}. \tag{1.4}$$

Planar projection forms a group [9] and has a number of invariants which will be discussed in chapters 3 and 5.

We can now use equation (1.3) to show that the above planar projective transformation can be approximated by an affine transformation when a planar object's

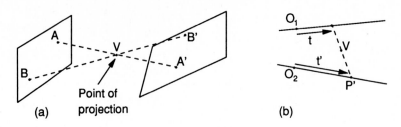

Figure 1.2: Planar and linear projection.

(a) Projection of points on one plane to points on another. (b) Projection of points on one line to points on another.

depth is small compared with its distance from the camera. Assume we have a set of coplanar points (X_i', Y_i', Z_i'), $i = 1, \ldots, N$, for which $Z_i' = Z' + \delta Z_i'$. The image coordinates $x_i' = X_i'/Z_i'$, $y_i' = Y_i'/Z_i'$ are approximately given by

$$x_i' = \frac{X_i'}{Z'}, \quad y_i' = \frac{Y_i'}{Z'} \quad \text{if} \quad Z' \gg \delta Z_i'.$$

Thompson & Mundy [11] use the rule of thumb that $(Z_{\max}' - Z_{\min}')/Z'$ should be less than 0.1 for the affine approximation to hold, with $Z_{\max}'$ being the largest value of Z_i' and $Z_{\min}'$ the smallest. If we let $s = Z'$, (1.3) becomes

$$\begin{bmatrix} X' \\ Y' \\ Z' \end{bmatrix} = \begin{bmatrix} a_{11} & a_{12} & a_{13} \\ a_{21} & a_{22} & a_{23} \\ 0 & 0 & s \end{bmatrix} \begin{bmatrix} X \\ Y \\ 1 \end{bmatrix}. \tag{1.5}$$

Defining the elements c_{ij} of the 3×3 matrix C_1 by $c_{ij} = a_{ij}/s$, so that (1.5) becomes $\mathbf{p}' = \mathbf{C_1 q}$, letting $\mathbf{C_2}$ define a second view and $\mathbf{D} = \mathbf{C_2 C_1^{-1}}$ allows one to show that the image coordinates $\hat{\mathbf{h}}^T = [\hat{x} \ \hat{y} \ 1]$ of the second view are related to those of the first view $\mathbf{h}' = [x' \ y' \ 1]$ by $\hat{\mathbf{h}} = \mathbf{D h}'$:

$$\begin{bmatrix} \hat{x} \\ \hat{y} \\ 1 \end{bmatrix} = \begin{bmatrix} d_{11} & d_{12} & d_{13} \\ d_{21} & d_{22} & d_{23} \\ 0 & 0 & 1 \end{bmatrix} \begin{bmatrix} x' \\ y' \\ 1 \end{bmatrix} \quad \text{— an affine transformation.}$$

This can alternatively be written as

$$\begin{bmatrix} \hat{x} \\ \hat{y} \end{bmatrix} = \begin{bmatrix} d_{11} & d_{12} \\ d_{21} & d_{22} \end{bmatrix} \begin{bmatrix} x' \\ y' \end{bmatrix} + \begin{bmatrix} d_{13} \\ d_{23} \end{bmatrix}. \tag{1.6}$$

Since the affine transformation is linear, invariant features are much easier to find than in the case of projective transformations. Examples of affine transformations are shown in figure 1.4. Note that parallel lines remain parallel under such transformations.

Just as a linear transformation in 3-D is equivalent to planar (2-D) projection, so a linear transformation in 2-D is equivalent to a 1-D projection of points on one line onto another line — see figure 1.2 (b).

$$\text{1-D projection:} \quad t' = \frac{\alpha t + \beta}{\gamma t + \delta}, \quad \alpha, \beta, \gamma, \delta \quad \text{real.}$$

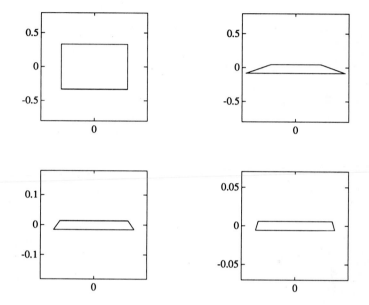

Figure 1.3: Perspective views of a rectangle.

The projective transformation tends towards an affine one as the object recedes into the distance. **Top:** *the rectangle's centre is 3 units from the point of projection; head on and at 80°.* **Bottom:** *12 and 30 units away.*

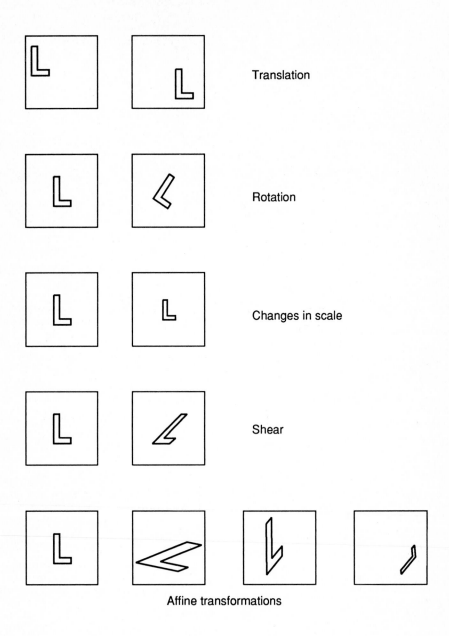

Figure 1.4: Examples of affine image transformations.

In homogeneous coordinates, with $t = x/y$ and $t' = x'/y'$, we have

$$\begin{bmatrix} x' \\ y' \end{bmatrix} = \begin{bmatrix} \alpha & \beta \\ \gamma & \delta \end{bmatrix} \begin{bmatrix} x \\ y \end{bmatrix}.$$

The 1D projective transformation has three degrees of freedom, so we can arbitrarily set $\delta \equiv 1$. Any three distinct collinear points can be projected to any other such triple of points under a 1D projective transformation.

Further approximations are possible under constrained viewing conditions; for instance, if an object is viewed from a fixed camera and is constrained to lie on a flat surface parallel to the image plane near its intersection with the optical (Z) axis (the *fronto-parallel* configuration), the affine approximation reduces to rotations, translations and scale changes. This is relevant if one would like to automate inspection of objects on a conveyor belt for example. An application in which one is only interested in rotation invariants is automatic fault detection in the heads of bolts, which slide head-up along two parallel guides into a fixed position relative to the camera, save for possible rotations.

Examples of invariant features in this case are the angle between two lines, which is unchanged by image translations, rotations and changes in scale, and the ratio of distances along a line or the ratio of two areas, both of which are invariant to affine transformations. Clearly, any function which is invariant to a given set of transformations is invariant to any subset of transformations — projective invariants are also invariant to affine transformations etc.

1.3 Overview

An object is said to be partially occluded when part of it is obscured from view by another object. Anyone glancing around a room will quickly see that partial occlusion is the norm rather than the exception. Our goal is to find robust techniques for recognizing partially occluded near-planar objects. To reach it, we must first obtain a thorough understanding of invariant features for unoccluded objects, which is the task of chapters 2 to 5. These chapters build up successively from invariance to simple transformations in chapter 2 to invariance to full planar projection in chapter 5: chapter 2 looks at invariance to translations, rotations, changes in scale, changes in contrast and combinations of these, first for ideal (continuous) images and then for discrete ones (as processed by computers); chapter 3 presents a tutorial introduction to the theory of algebraic invariants, which is fundamentally important when dealing with affine invariants, the subject of chapter 4, and projective invariants, the subject of chapter 5. Chapter 4 gives a definitive account of methods for obtaining invariance to affine transformations using *moments*, as well as summarizing alternative techniques and presenting a novel one based on *correlations*. Chapter 5 shows how to apply the results of chapter 3 to recognizing perspective views of objects, summarizes some of the authors results as well as discussing alternative methods. Finally, chapter 6 discusses schemes for using the invariant features of chapters 4 and 5 to recognize partially occluded objects irrespective of viewpoint, and chapter 7 provides a summary and conclusions.

To the author's knowledge, no monograph on the subject of using invariant features to recognize objects has appeared at the time of writing[1], with most results having appeared in conferences and workshops rather than journals (see in particular the collection of papers based on a workshop in 1991, reference [10]). Hence chapters 2 to 5, although presenting quite a number of new results along the way, are written to provide a comprehensive and up-to-date summary of research performed to date.

As mentioned above, chapters 2 to 5 are written to provide an understanding of invariant functions to aid in our goal of recognizing partially occluded objects. One important aspect is their robustness to image distortion. Many authors consider the effects of adding Gaussian noise to the image to test the robustness of invariants (e.g. [14, 15, 16]), but in most cases the amount of noise in the image is almost imperceptible; a more useful test of robustness is to see how the invariant features react to distortions of the object's shape, especially those caused by the discrete nature of the image when viewing distant or small objects. Theoretical results for rotation invariance are presented in chapter 2, including a novel family of invariant features, and experimental results are presented in both chapters 2 and 4; the main conclusion is that the so-called moment invariants are more robust than other invariants, contrary to beliefs voiced in the literature on the subject (see for example reference [17]).

When using image features to recognize objects, one would like them to provide good discrimination between different objects. In the case of image invariants, one is first interested in invariance to geometric transformations; having found invariant features, one must then investigate whether they provide discrimination. The experiments in chapters 2 and 4 also examine this; again, the moment invariants are shown to perform well.

If we are to recognize partially occluded objects, the information required for recognition must be available locally as well as globally, and the question naturally arises as to whether one can use the image invariants derived in chapters 2 to 5 to recognize partially occluded objects. The answer is in the affirmative, and is the subject of chapter 6. Numerous authors claim that moment invariants cannot be used under partial occlusion (see for example [18]), but this is not so: a number of schemes for their use are presented, and it is shown that they have significant advantages over other invariants, both in theory and in practice, the latter being demonstrated by experiment.

The next section ends this introductory chapter with a brief discussion of the image primitives that are used to generate invariant functions in chapters 4 to 6.

1.4   Image primitives

In order to compute invariant functions of an image, one often needs to extract certain features from the image, called primitives, which are then used in the construction of invariant functions. For instance, if one is using the angle between lines as an invariant then the lines are the primitives. As we will see in the following chapters, one can either compute invariant functions based on the intensity function of the image or based on the shape of the image boundary (in both cases one assumes that the object

---

[1]Wechsler's review [12] discusses some simple invariants and Kanatani's book [13] contains an extensive treatment of image invariants, but he only considers invariance to camera rotations and does not attempt to obtain invariance to general object motion relative to the camera. Furthermore, as discussed in chapter 5, some of his analysis is incorrect.

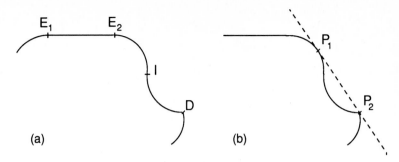

Figure 1.5: Finding reference points on a curve.

(a) E_i are end points of a line; I is a point of inflexion; D is a sharp discontinuity in curvature. Only D is a robust feature point in this case. (b) P_1 and P_2 are two points where the curve's convex hull leaves the curve itself; they are generally robust reference points, and are called bitangent points.

has been correctly separated from the background). Many methods, especially those that seek to recognize partially occluded objects, use points and lines as primitives; below we will briefly look at how these primitives can be extracted from the object's boundary.

The standard approach to finding reference points relies on differential properties of an object's boundary. Discontinuities in curvature and points of inflexion of planar curves are invariant under projection, but not always robust: the end point of a straight line is a discontinuity in curvature that is difficult to detect when the line goes into a curve of low curvature; similarly, points of inflexion can often be confused with a short line. A more robust technique uses *bitangent points* where two distinct points on the boundary share the same tangent (see figure 1.5) [9]. Also robust are points where a discontinuity in curvature is linked with a fairly sharp change in direction. These latter points are usually extracted by detecting large extrema of curvature — although extrema are theoretically not stable under projection, in practice they are [19, 20].

Further reference points can be obtained by noting where tangents constructed using the above points intersect with other tangents or with the boundary [9] (the latter being more robust against noise). As pointed out by Lamdan *et al.* [21], almost all near-planar objects have concavities, so this approach can be used in most cases. Forsyth *et al.* [9] present a generally robust method of extracting four reference points from a concavity that does not rely on extrema of curvature (figure 1.6(a)). The method fails when there is a discontinuity of curvature at points A or D such that the curvature has a different sign on either side of the discontinuity (point A in figure 1.6b), but in this case one can extract the tangents of the boundary on both sides of the discontinuity (figure 1.6b).

Points and lines will be used as image primitives in chapter 4 on affine invariants and in chapter 5 on projective invariants; although they can be used for the simpler transformations discussed in chapter 2, alternative global techniques will be used instead.

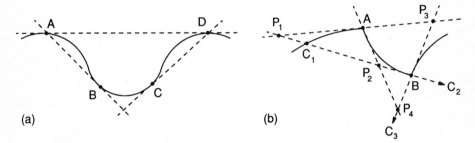

Figure 1.6: Using concavities to obtain projective invariants.

(a) A and D are points where the convex hull meets the boundary. B and C are the points where lines through A and D are tangent with the concave part of the boundary. (b) One can use the tangents at either side of a discontinuity in curvature to obtain a number of reference points. C_2 and C_3 are points where the tangents intersect with the boundary on the other side.

Chapter 2

Translation, Rotation, Scale and Contrast Invariants

2.1 Introduction

The image of a planar object that is constrained to lie on a flat surface parallel to the image plane and close enough to the optic axis to allow one to ignore perspective distortions (the fronto-parallel configuration) undergoes translation, scaling and rotation as the object moves on the surface. This chapter is devoted to features invariant to such transformations, as well as to changes in image brightness and contrast (see figure 2.1).

If the object is constrained to lie at a constant distance from the camera, invariance to changes in scale is not required; likewise, if one wants to distinguish between different orientations of an object, rotation invariance should be dropped, and if one wants sensitivity to translation, translation invariance should be dropped. An example of an application where one would only be interested in rotation invariance is the automatic inspection of bolt heads for faults, where each bolt is viewed in the same position, except rotations.

There are two approaches to forming features invariant to linear image transformations: one involves normalizing the image by finding the linear coordinate transformation that results in a standard version of the image which is invariant to the stipulated transformations; the other involves finding invariant functions of the image directly, without performing a coordinate transformation. The latter will be dealt with first, while the former is discussed in section 2.9.

If one limits oneself to binary images, one can form invariants using the derivatives of the object's boundary [22]; however, they are less robust than the features discussed below, and will not be discussed further. Another technique limited to binary images uses Fourier descriptors [4]. If one considers the radius of the boundary from the centroid at a given angle θ as a function $r(\theta)$ then $r(\theta)$ is periodic and can be described using a complex Fourier series. The magnitude of the coefficients of this series are the Fourier descriptors of the boundary, and are invariant to rotations and translations; invariance to scale can be obtained by using ratios of the Fourier coefficients [4]. Since these features cannot be used to discriminate between objects with identical boundaries (e.g. books with different covers), they are not considered further. Nevertheless, experiments indicate that they perform well on binary images — see the articles by Reeves *et al.* [4] and by Chen & Ho [5].

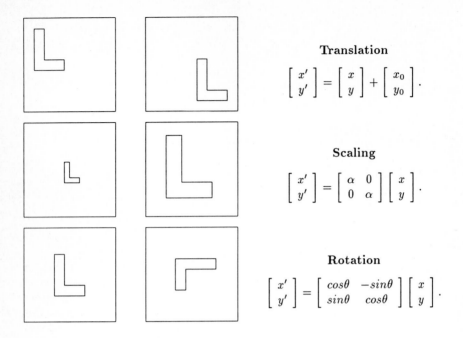

Translation

$$\left[\begin{array}{c} x' \\ y' \end{array} \right] = \left[\begin{array}{c} x \\ y \end{array} \right] + \left[\begin{array}{c} x_0 \\ y_0 \end{array} \right].$$

Scaling

$$\left[\begin{array}{c} x' \\ y' \end{array} \right] = \left[\begin{array}{cc} \alpha & 0 \\ 0 & \alpha \end{array} \right] \left[\begin{array}{c} x \\ y \end{array} \right].$$

Rotation

$$\left[\begin{array}{c} x' \\ y' \end{array} \right] = \left[\begin{array}{cc} cos\theta & -sin\theta \\ sin\theta & cos\theta \end{array} \right] \left[\begin{array}{c} x \\ y \end{array} \right].$$

Changes in brightness and contrast: $f'(x,y) = cf(x,y) + b.$

Figure 2.1: The transformations dealt with in this chapter.

The features discussed below are not limited to binary images. However, if grey level images are used, care must be taken to ensure that the surfaces reflect light uniformly (e.g. Lambertian). In all cases it is assumed that the object of interest has been extracted from its surroundings and appears on a background with zero intensity.

As we shall see, invariance to image translations, rotations and changes of scale can be dealt with either using image correlations or using image moments. Of the latter, a novel set of orthogonal moments, called the real weighted Fourier moments, are introduced and shown to perform better than conventional moments.

The chapter is in two parts: the first part, sections 2.2 to 2.9, discusses invariants of ideal (continuous) images, while the second part discusses the effect of using discrete images instead of continuous ones. The first part starts by looking at features invariant to translations, then moves on to rotations, changes in scale, combinations of these and then features invariant to changes in contrast; it concludes with a discussion of how to obtain invariance to the above by *normalizing* the image. The second part starts with an analysis of rotation invariance for discrete images: under what circumstances can one form absolute invariants, and are the features that were orthogonal in the continuous case still orthogonal in the discrete one? Next, it is shown how to modify correlation invariants to deal with discrete images. Finally the invariants' computational cost and performance in noise is analysed and results of experiments that test the invariants' robustness to the distortions in shape caused by coarse sampling are presented.

Many of the concepts introduced below will be used in the chapter on affine invariants; these are the moments and correlations defined in section 2.2 on translation invariance (the moments are also used in chapter 6), the complex moments defined in section 2.4.2 on rotation invariance, contrast and brightness invariance (section 2.8), normalization (section 2.9) and the discussion of performance and computational cost in section 2.12.

2.2 Translation invariants

In the first half of this section image correlations will be defined and shown to be invariant to translations of the image. Image moments are defined in the second half, as are the *central moments* which are also shown to be invariant to image translations.

The image intensity distribution is represented by the continuous function $f(x, y)$, which gives the intensity of the image at the point (x, y) in the image. Furthermore, the image $f(x, y)$ is assumed to have finite support (i.e. the area for which $f(x, y) \neq 0$ is finite) and to be non-negative everywhere. In practice images are discrete, in which case the integrals are replaced by summations; this will be discussed at length in section 2.10.

2.2.1 Image correlations

Below the first, second and third order correlations $g_1, g_2(\mathbf{a}), g_3(\mathbf{a}_1, \mathbf{a}_2)$ of the image $f(x, y)$ are defined [23], followed by the general definition of the kth order correlation. $\mathbf{a}_i^T = [\alpha_i \ \beta_i]$ is a 2-D vector; a point (x, y) on the image is represented by $f(\mathbf{x})$, where $\mathbf{x}^T = [x \ y]$; the integrals are over the range $[-\infty, +\infty]$:

$$g_1 = \int f(\mathbf{x}) \, d\mathbf{x};$$

$$g_2(\mathbf{a}) = \int f(\mathbf{x}) f(\mathbf{x} + \mathbf{a}) \, d\mathbf{x};$$

$$g_3(\mathbf{a}_1, \mathbf{a}_2) = \int f(\mathbf{x}) f(\mathbf{x} + \mathbf{a}_1) f(\mathbf{x} + \mathbf{a}_2) \, d\mathbf{x}.$$

The kth order correlation is given by

$$g_k(\mathbf{a}_1, \ldots, \mathbf{a}_{k-1}) = \int f(\mathbf{x}) f(\mathbf{x} + \mathbf{a}_1) \cdots f(\mathbf{x} + \mathbf{a}_{k-1}) \, d\mathbf{x}. \tag{2.1}$$

Since the integrals are over the whole range of $\mathbf{x}$, translating the image by an amount $\mathbf{x}_0$ will leave the values of the correlations unaffected; in other words they are translation invariant. It can easily be shown that the Fourier transform of the second order correlations equals the Fourier power spectrum of the image $f(x, y)$; hence we can use the magnitude of the image's Fourier transform as translation invariants instead of the second order correlations. The Fourier transform $F(u, v)$ of an image $f(x, y)$ is defined as

$$F(u, v) = \int_{-\infty}^{+\infty} \int_{-\infty}^{+\infty} f(x, y) e^{-2\pi j (ux + vy)} \, dx \, dy$$

where $j^2 \equiv -1$. $|F(u,v)|$ is translation invariant.

The first and second order correlations are not complete representations of the image, which means that information is lost, allowing many different images to have the same first and second order correlations. Third order correlations of discrete binary images are complete representations (proved by Minsky & Papert [24]), but are computationally intensive since they are four dimensional. One can think of the kth order correlation integral as defining a mapping from the continuous 2-D image $f(x,y)$ to the continuous 2^{k-1} dimensional correlation function $g_k(.)$; in the case of 1st and 2nd order correlations it is a many-to-one mapping.

2.2.2   Moment invariants

Moment invariants were first introduced by Hu [25, 26] based on his fundamental theorem of moment invariants. It turns out that this theorem, which has recently been corrected by the author [27], is only needed to derive invariants to affine transformations; a discussion will be left to chapter 4. Moments are used extensively in this chapter and also in chapters 4 and 6.

The *regular moment* m_{pq} of an image $f(x,y)$ is defined as

$$m_{pq} = \int_{-\infty}^{+\infty} \int_{-\infty}^{+\infty} x^p y^q f(x,y) \, dx \, dy, \qquad p, \, q = 0, \, 1, \, 2, \cdots. \qquad (2.2)$$

If the function $f(x,y)$ is piecewise continuous in addition to having compact support one can prove that moments of all orders $p+q$ exist, that the infinite set of moments uniquely determines $f(x,y)$ and conversely that they are themselves uniquely determined by $f(x,y)$. This means that the integral (2.2) can be seen as a one to one mapping of the continuous, finite area image $f(x,y)$ onto the infinite discrete moment matrix $\mathbf{M}$ with entries m_{pq}.

The *central moments* μ_{pq} are defined as

$$\mu_{pq} = \int_{-\infty}^{+\infty} \int_{-\infty}^{+\infty} (x - \bar{x})^p (y - \bar{y})^q f(x,y) \, dx \, dy, \qquad p, \, q = 0, \, 1, \, 2, \cdots, \qquad (2.3)$$

where $\bar{x} = m_{10}/m_{00}$ and $\bar{y} = m_{01}/m_{00}$. The central moments are equivalent to the regular moments of an image that has been shifted so that the image centroid $(\bar{x}, \bar{y})$ coincides with the origin; as a result central moments are invariant to image translations, and $\mu_{10} \equiv \mu_{01} \equiv 0$. The central moments are related to the regular moments by

$$\mu_{pq} = \sum_{k=0}^{p} \sum_{l=0}^{q} \binom{p}{k} \binom{q}{l} (-1)^{k-l} m_{p-k,q-l} m_{10}^{k} m_{01}^{l} m_{00}^{-(k+l)}, \qquad (2.4)$$

where $\qquad \binom{p}{k} = \dfrac{p!}{k! \, (p-k)!}.$

2.3   The complex-logarithmic mapping

The complex-logarithmic mapping is a coordinate transformation that maps a function $f(x,y)$ in the input space to a function $g(\rho, \theta)$ in the output space such that

$$\rho = \ln r = \frac{1}{2} \ln(x^2 + y^2), \qquad \theta = \arctan\left(\frac{y}{x}\right).$$

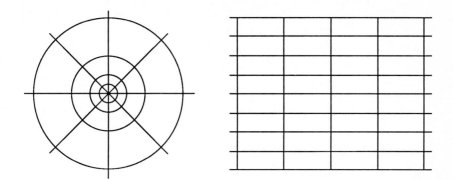

Figure 2.2: The complex-logarithmic mapping.

The complex-logarithmic mapping is a conformal mapping so angles between lines are preserved. The concentric circles on the image (left) maps to vertical lines on the mapped image (right), while the radial lines on the image map to horizontal lines. The shape on the right in the mapped image corresponds to the larger square in the image.

Image rotations are transformed to 2π periodic translations along the θ-axis and changes of scale to translations along the ρ-axis — see Fig. 2.2. Casasent & Psaltis [28] were the first to use this mapping to obtain invariances in image recognition. Problems arise from the singularity at the origin, forcing one to discard a small central disc in the image, and causing stretching problems when objects are enlarged [29]: if, as in Fig. 2.2, a small square of uniform brightness is enlarged, the edges translate along the ρ-axis, which means that the whole object is stretched along the ρ-axis. The only way that this stretching will not cause problems is if the image has zero intensity in the region around the origin. Wechsler [29] suggests pre-processing the image to extract edges before performing the mapping as a means of achieving this; however, this only succeeds if the edges are not close to the origin.

The complex-logarithmic mapping has some interesting features which are worth mentioning briefly. First, there is strong physiological evidence [30, 31, 32] that the mapping, or one very similar to it, is implemented in our brains. Second, the complex-logarithmic mapping is conformal and so can be justified from information-theoretic arguments. Conformal mappings preserve angles between corresponding pairs of lines in the input and output spaces. The psychologist Attneave showed that contours with high curvature are subjectively more important than those with low curvature [33]; Resnikoff [33] has justified this mathematically by showing that the angle α between two lines contains $I = \log_2(2\pi/\alpha)$ bits of information, and hence that smaller angles (sharper corners) contain more information. From the above it would seem that a mapping that preserves angles between lines, such as the complex-logarithmic one, could have advantages over non-conformal mappings.

2.4   Rotation invariants

Rotations of the image correspond to coordinate transformations of the form

$$\begin{bmatrix} x' \\ y' \end{bmatrix} = \begin{bmatrix} \cos\theta & -\sin\theta \\ \sin\theta & \cos\theta \end{bmatrix} \begin{bmatrix} x \\ y \end{bmatrix}.$$

Group theory tells us that the only one-dimensional invariants for 2-D rotations are given by the magnitude of $h_k(r)$ [34], the circular Fourier transform of the image $f(r, \theta)$ represented in polar coordinates:

$$h_k(r) = \int_0^{2\pi} f(r, \theta) e^{jk\theta} \, d\theta, \qquad k \text{ integer.}$$

Below we will first see how this relates to correlation invariants and then how it relates to moment invariants.

2.4.1   Correlations

The second order angular correlation of the image $f(r, \theta)$ in polar coordinates is

$$g_2(\alpha, r) = \int_0^{2\pi} f(r, \theta) f(r, \theta + \alpha) \, d\theta;$$

the extension to higher order correlations is obvious from (2.1). In a similar way to the correlations of section 2.2.1, it can easily be shown that the circular Fourier transform of $g_2(\alpha, r)$ equals the squared magnitude of $h_k(r)$. As before one can use higher order correlations to avoid losing information. Angular correlations of discrete images are not a natural extension of the continuous case because of the varying number of pixels at a given radius; how to apply correlations to such images is discussed in section 2.11.

2.4.2   Moments

The first four of Hu's translation and rotation invariants are given by [25]:

$$\begin{align}
\phi_1 &= \mu_{20} + \mu_{02}, \tag{2.5} \\
\phi_2 &= (\mu_{20} - \mu_{02})^2 + 4\mu_{11}^2, \\
\phi_3 &= (\mu_{30} - 3\mu_{12})^2 + (3\mu_{21} - \mu_{03})^2, \\
\phi_4 &= (\mu_{30} + \mu_{12})^2 + (\mu_{21} + \mu_{03})^2. \tag{2.6}
\end{align}$$

These can most easily be derived using complex moments c_{pq}, which were introduced by Davis [35] and re-introduced by Abu-Mostafa & Psaltis [36, 14]; they are defined as:

$$c_{pq} = \int_{-\infty}^{+\infty} \int_{-\infty}^{+\infty} (x + jy)^p (x - jy)^q \, dx \, dy, \tag{2.7}$$

where $j^2 \equiv -1$. If we map to polar coordinates, so that $x + jy = re^{j\theta}$, (2.7) becomes

$$c_{pq} = \int_0^{2\pi} \int_0^{\infty} r^{p+q+1} e^{j(p-q)\theta} f(r, \theta) \, dr \, d\theta$$

$$= \int_0^\infty r^{p+q+1} \left\{ \int_0^{2\pi} f(r,\theta) e^{jk\theta} \, d\theta \right\} dr, \quad k = p - q,$$

$$= \int_0^\infty r^{p+q+1} h_k(r) \, dr. \tag{2.8}$$

From this we see that $|c_{pq}|$ is invariant to rotations of $f(r,\theta)$, and that c_{pq} is the projection of the image's circular Fourier transform $h_{p-q}(r)$ onto the function r^{p+q+1}. In other words the continuous one dimensional function $h_{p-q}(r)$ is mapped onto an infinite vector with elements c_{pq}. If an image has n-fold rotational symmetry, all c_{pq}'s for which $p - q$ is not divisible by n are identically zero [14] — for example, 'I' has twofold rotational symmetry, an equilateral triangle threefold, a square fourfold etc.

It is clear from (2.7) and the definition of moments that, for a centred image,

$$\begin{aligned} c_{11} &= \mu_{20} + j\mu_{02}, \\ c_{21} &= (\mu_{30} + \mu_{12})^2 + j(\mu_{21} + \mu_{03})^2, \quad \text{etc.} \end{aligned}$$

It is also apparent from (2.8) that using regular or central moments limits one to using radial moments of integer order $p + q + 1$. If one is interested in features invariant to rotation but not to translations, which would be the case if the images were normalized to be invariant to translations, the restriction on the powers of r can be lifted; indeed, one can use any radial weighting function, since the magnitude of $h_k(r)$ is invariant. This is discussed further in the next section.

2.4.3   Orthogonal moments

Teague [37] introduced the idea of orthogonal moments for rotation invariance. Moments as defined in (2.2) can be seen as the projection of $f(x,y)$ onto the non-orthogonal basis $\{x^p y^q\}$; as a result these moments are correlated with one another. Teague considered images defined on a disc of unit radius and used the orthogonal Zernike polynomials as a basis on which to project the image. The Zernike polynomials $V_{pq}(x,y)$ are the only polynomials in x and y that are orthogonal over the unit disc D for which $x^2 + y^2 \le 1$ [38], i.e.

$$\iint_D V_{pq}(x,y) V_{lm}(x,y) \, dx \, dy = \delta_{pl} \delta_{qm}. \tag{2.9}$$

The function must be a polynomial in x and y to obtain rotation *and* translation invariants, since only projections of the image $f(x,y)$ onto such polynomials result in functions of the regular moments; translation invariance is achieved by replacing the regular moments in these functions by the central moments. Teh & Chin [15] and Khotanzad *et al.* [39, 16] perform experiments using Zernike moments and show that they perform well in practice.

As before, if we choose to ignore translation invariance, we are free to choose from an infinite number of orthogonal functions. For example, transforming to polar coordinates and writing the basis as $v_{pq}(r,\theta)$, (2.9) becomes

$$\int_0^{2\pi} \int_0^\infty v_{p'q'}^*(r,\theta) v_{pq}(r,\theta) \, r \, dr \, d\theta = \delta_{p'p} \delta_{q'q}.$$

If the basis is of the form $v_{pq}(r,\theta) = (1/2\pi).h_p(r)e^{jq\theta}$, then we get

$$\int_0^\infty r h_{p'}^*(r) h_p(r) \left\{ \frac{1}{2\pi} \int_0^{2\pi} e^{j(q-q')\theta}\, d\theta \right\} dr \; = \; \delta_{p'p}\delta_{q'q}.$$

The term in brackets equals $\delta_{q'q}$, so we see that the condition for orthogonality reduces to

$$\int_0^\infty r h_{p'}^*(r) h_p(r)\, dr \; = \; \delta_{p'p}. \tag{2.10}$$

Hence, any such polar separable function will be orthogonal over the unit disc if $\sqrt{r}R(r)$ is orthogonal over the interval $[0, 1]$; the Zernike polynomials are obtained by orthogonalising $\sqrt{r} \times \{r^{|l|}, r^{|l|+2}, r^{|l|+4}, \ldots\}$ over the interval $[0, 1]$, and the pseudo-Zernike polynomials are obtained by orthogonalising $\sqrt{r} \times \{r^{|l|}, r^{|l|+1}, r^{|l|+2}, \ldots\}$ [38, 15]. The radial function r^p increasingly emphasises portions of the image disc near the outer rim as p increases; Abu-Mostafa & Psaltis [36] attribute the increasing noise sensitivity of moments as the order increases to this, and Teh & Chin [15] show that a given number of pseudo-Zernike moments are nearly an order of magnitude less sensitive to additive white noise than the same number of Zernike moments. If we set l to zero, we see that the former result from orthogonalising $\sqrt{r} \times \{1, r, r^2, \ldots\}$ and the latter from orthogonalising $\sqrt{r} \times \{1, r^2, r^4, \ldots\}$. It is therefore possible that one could achieve a further improvement in robustness against noise by orthogonalising $\sqrt{r} \times \{1, r^{1/N}, r^{2/N}, \ldots\}$, $N > 1$. This can easily be done using Jacobi polynomials [38] and results in a radial polynomial that is the radial part of what I call the general-Zernike polynomial:

$$R_{k/N}(r) = \sqrt{2(1 + \frac{k}{N})} \sum_{s=0}^k (-1)^s \frac{(2N+k+s-1)!}{(k-s)!\, s!\, (2N+s-1)!} r^{s/N}.$$

Zernike polynomials correspond to $N = \frac{1}{2}$, pseudo-Zernike polynomials to $N = 1$. The general-Zernike basis is then

$$V_{kl/N}(r,\theta) = \frac{1}{\sqrt{2\pi}} R_{k/N}(r)e^{jl\theta},$$

and the corresponding general-Zernike moments of an image $f(r,\theta)$ are defined by

$$Z_{kl/N} = \int_0^{2\pi} \int_0^1 V_{kl/N}^*(r,\theta) f(r,\theta) r\, dr\, d\theta, \tag{2.11}$$

where V^* is the complex-conjugate of V.

Figures. 2.3 – 2.6 show plots of the functions $\sqrt{r}R_{k/N}(r)$ orthogonal over $r \in [0, 1]$, for N = 0.5, 1, 2 and 3; they show that the pseudo-Zernike polynomials weighted by $\sqrt{r}$ span the interval $[0, 1]$ more uniformly than the others, and are hence likely to be the most robust.

The plots indicate that the frequency varies with r; instead of using a polynomial basis, it seems natural to try a basis made up of sines and cosines, so the frequency of each basis function is constant. The real weighted Fourier (RWF) basis is defined as:

$$\begin{aligned} h_{p1}(r) &= \tfrac{1}{\sqrt{r}} \cos 2p\pi r; \\ h_{p2}(r) &= \tfrac{1}{\sqrt{r}} \sin 2p\pi r; \end{aligned} \qquad p \text{ integer.}$$

Experimental results are presented in the next section, where we will see that the RWF moments outperform the pseudo-Zernike moments.

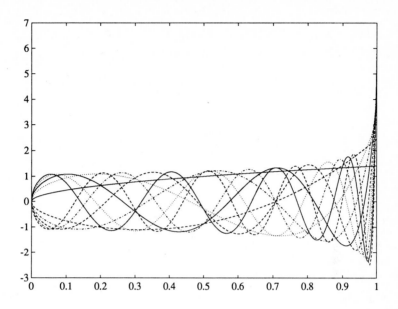

Figure 2.3: The first ten weighted radial general-Zernike polynomials with N = 0.5 (i.e. Zernike polynomial).

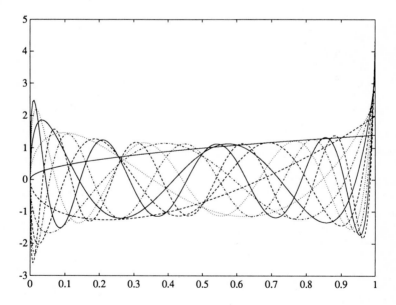

Figure 2.4: The first ten weighted radial general-Zernike polynomials with N = 1 (i.e. pseudo-Zernike polynomial).

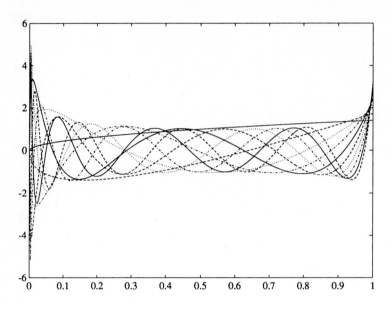

Figure 2.5: The first ten weighted radial general-Zernike polynomials with N = 2.

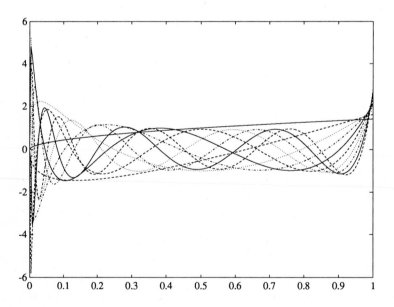

Figure 2.6: The first ten weighted radial general-Zernike polynomials with N = 3.

2.5 Scale invariants

Scale changes are caused by coordinate transformations of the form

$$\begin{bmatrix} x' \\ y' \end{bmatrix} = \begin{bmatrix} \alpha & 0 \\ 0 & \alpha \end{bmatrix} \begin{bmatrix} x \\ y \end{bmatrix}. \tag{2.12}$$

Scale invariants can either be obtained by using correlations of the logarithmically transformed image, or by using moments by dividing each moment by a normalizing function that cancels out the effect of scaling.

2.5.1 Correlations

Transforming the x- and y-axes by taking their logarithm to obtain new axes $x' = \ln x$, $y' = \ln y$, where $x, y > 0$, results in scale changes becoming translations, but with the stretching problems mentioned in section 2.3. If we let $f'(x', y')$ be the image after the mapping of the axes, so $f'(x', y') = f'(\ln x, \ln y) = f(x, y)$, then the second order correlations in the new axes can be expressed in terms of the original axes as follows:

$$g_2(\alpha_1', \alpha_2') = \int_0^\infty \int_0^\infty f'(x', y') f'(x' + \alpha_1', y' + \alpha_2') \, dx' \, dy'.$$

Let $\alpha_i' = \ln \alpha_i$, so $x' + \alpha_1' = \ln x + \ln \alpha_1 = \ln(\alpha_1 x)$, and

$$g_2(\ln \alpha_1, \ln \alpha_2) = \int_0^\infty \int_0^\infty f(x, y) f(\alpha_1 x, \alpha_2 y) \frac{dx}{x} \frac{dy}{y}.$$

This can easily be generalized to higher-order correlations.

As before, the second order correlations are equal to the squared magnitude of the Fourier transform of the image in the new coordinates; when written in terms of the original coordinates this Fourier transform is known as the Mellin transform and is given by

$$\begin{aligned} M(u, v) &= \int_{-\infty}^{+\infty} \int_{-\infty}^{+\infty} f'(x', y') e^{-j(ux' + vy')} \, dx' \, dy' \\ &= \int_{-\infty}^{+\infty} \int_{-\infty}^{+\infty} f(x, y) x^{-ju-1} y^{-jv-1} \, dx \, dy. \end{aligned}$$

Scale-invariant correlation-based features of discrete images are discussed in section 2.11.

2.5.2 Moments

If we let $f'(x', y')$ represent the image $f(x, y)$ after scaling each axis by α (see (2.12)), so $f'(x', y') = f'(\alpha x, \alpha y) = f(x, y)$, and $x' = \alpha x$, $y' = \alpha y$, then we have

$$\begin{aligned} m_{pq}' &= \int_{-\infty}^{+\infty} \int_{-\infty}^{+\infty} x'^p y'^q f'(x', y') \, dx' \, dy' \\ &= \alpha^{p+q+2} \int_{-\infty}^{+\infty} \int_{-\infty}^{+\infty} x^p y^q f(x, y) \, dx \, dy. \end{aligned}$$

Hence $m'_{pq} = \alpha^{p+q+2}m_{pq}$, and similarly $\mu'_{pq} = \alpha^{p+q+2}\mu_{pq}$. This gives $\mu'_{00} = \alpha^2\mu_{00}$, from which we can see that

$$\eta_{pq} = \frac{\mu_{pq}}{\mu_{00}^{\gamma}}, \qquad \gamma = \frac{p+q}{2} + 1, \qquad p+q = 2, 3, \ldots,$$

is invariant to changes of scale [26, 40]:

$$\eta'_{pq} = \frac{\mu'_{pq}}{\mu_{00}^{\prime\gamma}} = \frac{\alpha^{p+q+2}\mu_{pq}}{\alpha^{2\gamma}\mu_{00}^{\gamma}} = \frac{\mu_{pq}}{\mu_{00}^{\gamma}} = \eta_{pq}.$$

2.6   Rotation and scale invariants

Correlation and moment based techniques achieve rotation and scale invariance by combining the two individual invariant functions.

2.6.1   Correlations

Invariance to both rotation and scale changes is achieved by computing angular and radial correlations after performing the complex-logarithmic mapping. If we represent the image in polar coordinates, the second order angular-radial correlation is

$$g(\alpha, \beta) = \int_0^{2\pi} \int_0^{\infty} f(r, \theta)f(re^{\alpha}, \theta + \beta)\, \frac{dr}{r}\, d\theta.$$

As always, higher order correlations can easily be defined. The Fourier transform of $g(\alpha, \beta)$ is the same as the squared magnitude of the radial-Mellin-circular-Fourier transform defined by

$$M(u, k) = \int_0^{2\pi} \int_0^{\infty} f(r, \theta)r^{-ju-1}e^{-jk\theta}\, dr\, d\theta, \qquad k \text{ integer}.$$

Two papers co-authored by Sheng [41, 42] have investigated a hybrid technique, which they misleadingly call Fourier-Mellin descriptors, that simply involves computing the radial moments of the circular Fourier transform of the image and normalizing them in the same way that was outlined in section 2.10.2.

2.6.2   Moments

Replacing central moments in the expressions for rotation invariants (2.5) and (2.6) by the scale invariants η_{pq} results in translation, rotation and scale invariants; the first two become:

$$\begin{aligned}
\psi_1 &= \eta_{20} + \eta_{02}; \\
\psi_2 &= (\eta_{20} - \eta_{02})^2 + 4\eta_{11}^2.
\end{aligned}$$

To drop the translation invariance one simply uses regular moments in defining η_{pq} rather than central moments.

2.7   Translation, rotation and scale invariants

We have already seen in section 2.6.2 that invariants to all three of the above transformations are easy to obtain using central moments. They pose much greater problems for correlations. The typical approach, first introduced by Casasent & Psaltis [28], involves first turning the image into a 2-D correlation function $g(\mathbf{a})$, which is translation invariant, then performing a complex-logarithmic mapping on $g(\mathbf{a})$ before finally computing the radial and angular correlations (here $\mathbf{a}^T = [u\ v]$):

$$g(u,v) = \int_{-\infty}^{+\infty} \int_{-\infty}^{+\infty} f(x,y)f(x+u,y+v)\,dx\,dy$$

$$h(\alpha,\beta) = \int_0^{2\pi} \int_0^\infty g(r,\theta)g(e^\alpha r, \theta+\beta)\,\frac{dr}{r}\,d\theta, \qquad (2.13)$$

where $g(r,\theta)$ is $g(u,v)$ in polar coordinates. Although Casasent & Psaltis claim that $h(\alpha,\beta)$ is invariant to all three transformations, this is not actually true. Consider how rotating and scaling the original image affects $g(u,v)$: rotation of the original image by an angle ϕ results in a rotation of $g(u,v)$ by the same amount; scaling the original image by a factor s, so the new coordinates x', y' satisfy $x' = sx$, $y' = sy$, results in $g(u,v)$ being scaled by $1/s$ and being multiplied by s^2: if $g'(u,v)$ is the correlation of the scaled image, then

$$g'(u,v) = s^2 g(\frac{u}{s}, \frac{v}{s}).$$

From this we see that $h'(\alpha,\beta) = s^4 h(\alpha,\beta)$. If one uses template matching to recognise images, as Casasent & Psaltis do, then the factor s^4 does not affect the outcome; otherwise one must use division to remove the factor as discussed in the next section.

Casasent & Psaltis used Fourier transforms performed optically in their scheme and implemented the logarithmic mapping using a cathode-ray tube. A drawback to this approach in addition to the stretching problems associated with the logarithmic mapping and the factor s^4, is that it loses even more information than just using second order correlations once: two images with different power spectra can result in identical invariants.

Fuchs & Haken [43, 44, 45, 46] have also used this method for their associative memory with apparently good results; however, the transformations are computationally expensive when not performed optically.

2.8   Contrast and brightness invariants

A change in contrast results in a new image intensity distribution $f'(x,y) = cf(x,y)$ whereas a change in brightness results in $f'(x,y) = f(x,y) + b$. In real images a flat or uniform change of brightness and contrast will only occur if the surface is Lambertian (reflects light in all directions), if the lighting is very homogeneous, or if the object is viewed at an angle from the same direction as the light source. A number of authors only consider changes in contrast [40, 36], assuming that any change in brightness has been compensated for or can be ignored. First we will discuss how to obtain invariants under this assumption; then, we will look at how to compensate for changes in brightness and in contrast using normalization.

Other than by normalization, invariants to changes in contrast can only be obtained by dividing out the factor introduced in the previous invariants; for instance, it follows from the definition of moments that the moments of $f'(x,y)$, m'_{pq}, are related to those of $f(x,y)$ by $m'_{pq} = cm_{pq}$, so m_{pq}/m_{00} is invariant to changes in contrast. Similarly the kth order correlation of $f'(x,y)$, $g'(\mathbf{a}_1, \ldots, \mathbf{a}_{k-1})$, is related to that of $f(x,y)$ by $g'(.) = c^k g(.)$, so $g(\mathbf{a}_1, \ldots, \mathbf{a}_{k-1})/[g(\mathbf{0}, \ldots, \mathbf{0})]^k$ is an example of a contrast invariant.

The above technique is no longer effective if the image intensity is also uniformly changed by an amount b i.e. $f'(x,y) = cf(x,y) + b$. Probably the most robust way of normalizing against changes in brightness b is to produce a normalized image $\hat{f}(x,y)$ that has zero mean intensity: $\hat{f}(x,y) = f(x,y) - \bar{f}$. If one is using moments to find the centroid one must beware, since m_{00} is equal to the mean intensity and is hence zero for the brightness-normalized image; instead of using the moments of the grey-level image, one can use the moments of the object's shape i.e. set all object pixels to intensity 1 and find the centroid using m_{00}, m_{10} and m_{01} of this binary version. This is the approach used in the experiment on affine invariance of grey-level images in chapter 4, section 4.5.3.

In addition to normalizing against changes in brightness, one can also obtain invariance to changes in contrast by normalizing the standard deviation of the object's pixel values. This is used to good effect in the experiment in chapter 4, section 4.5.3; as we will see, the conclusions are that it is essential to take both brightness and contrast changes into account, and that the above normalization is very effective.

2.9 Normalization using moments

Hu [26] was the first person to discuss image normalization; he introduced the principal axis of an image as a means of normalizing against rotations — the normalized image is obtained by rotating the input image so that its principal axis has a fixed, pre-determined orientation.

We will first see how the complex moments [35, 36] can be used to normalize an image to be invariant to translations, rotations and changes in scale and contrast, before looking at how to use the parameters of the normalizing transformation to compute the moments of the normalized image without having to interpolate, as first suggested by Udagawa [47, 4].

2.9.1 Normalizing against translations, rotations and/or reflections

Hu [26] introduced the concept of normalizing images against translations and rotations using moments. If the image's regular moments are denoted by m_{pq} and the central moments by μ_{pq} (see chapter 2), the image is normalized against translations by moving the origin to the point $(\bar{x}, \bar{y})$, where $\bar{x} = m_{10}/m_{00}$ and $\bar{y} = m_{01}/m_{00}$. Normalizing the image against rotations using moments is best done using complex moments [35, 14]; these are defined in equation (2.7) and repeated here for convenience:

$$c_{pq} = \int_{-\infty}^{+\infty} \int_{-\infty}^{+\infty} (x + jy)^p (x - jy)^q f(x,y) \, dx \, dy, \tag{2.14}$$

where $j^2 \equiv -1$. If we map to polar coordinates so that $x + jy = re^{j\theta}$, (2.14) becomes

$$c_{pq} = \int_0^{2\pi} \int_0^\infty r^{p+q+1} f(r, \theta) e^{j(p-q)\theta} \, dr \, d\theta.$$

From this it is clear that if we rotate the axes counterclockwise by an angle ϕ, the complex moment c'_{pq} of the new image is related to that of the original image by

$$c'_{pq} = c_{pq} e^{-j(p-q)\phi}. \tag{2.15}$$

If $p - q = 1$, there will be one unique angle $\phi \in [0, 2\pi]$ that results in c'_{pq} being positive real. The image can thus be normalized against rotations by rotating it clockwise by an amount ϕ. (In general, if $p - q = n$, there will be n distinct angles that result in c'_{pq} being positive real.) Problems arise if the image has n-fold rotational symmetry with $n > 1$, since in this case all c_{pq}'s for which $p - q$ is not divisible by n are identically zero [14]. To normalize images with rotational symmetry one must find the smallest value of $p - q$ for which $c_{pq} \neq 0$ and rotate the image clockwise by ϕ, which is obtained from (2.15). Recently the concept of a principal axis has been generalized so as to apply to objects with rotational symmetries [48]; however, normalization using these generalized axes is essentially no different from using the complex moments.

Invariance to reflections is important if one is classifying the shape of planar objects in 3-D, since one could be looking at them from in front or from behind; the latter would result in a reflection compared to the former. A reflection of the image about the x-axis results in a new image whose complex moments c'_{pq} are the complex conjugate of those of the original image: $c'_{pq} = c^*_{pq}$. Now, we saw above that normalizing the image against rotations results in our chosen c_{pq} being positive real; since this means that $c^*_{pq} = c_{pq}$, we see that after normalization against rotation, any reflection will be about the x-axis: only such a reflection leaves c_{pq} unaffected. To normalize against rotation and reflection, first normalize against rotation using c_{pq} and then, if the imaginary part of some other complex moment c_{mn} is negative, reflect the rotation-normalized image about the x-axis.

The fact that $c'_{pq} = c^*_{pq}$ allows one to test for reflectional symmetry very easily: all the complex moments of a normalized image with reflectional symmetry must be real. This is analogous to the skew invariants which will be discussed in chapter 4 being zero for images with a reflectional symmetry.

2.9.2 Normalizing against changes in scale and contrast

A change of scale by α results in a new image with moments m'_{pq} related to those of the old image by $m'_{pq} = \alpha^{p+q+2} m_{pq}$ [26, 27]; similarly a change of contrast by a factor c results in $m'_{pq} = cm_{pq}$ [40, 27]. We have two parameters α and c, so we need two relations in the moments to define a normalized image (see [49] for an interesting discussion of this idea). The complex moments c_{kk} are real and unaffected by image rotations; $c_{00} = m_{00}$, $c_{11} = m_{20} + m_{02}$. Transforming the image so that $c_{00} = \beta_1$, $c_{11} = \beta_2$, β_i fixed, will result in a normalized image that is invariant to changes in scale and contrast.

2.9.3 Normalizing against translation, rotation and changes of scale and contrast

Abu-Mostafa & Psaltis [14] achieve the above normalization by setting $c_{00} = \alpha$, $c_{10} = 0$, $c_{11} = \beta$ and c_{pq} positive real for some $p > q$, $(p,q) \neq (1,0)$. Now, $c_{00} = m_{00}$; $c_{10} = m_{10} + jm_{01}$ and $c_{11} = m_{20} + m_{02}$. Hence, the two conditions $m_{10} = 0$ and $m_{01} = 0$ embodied in $c_{10} = 0$ result in normalization against translation; the two conditions $c_{00} = \alpha$ and $c_{11} = \beta$ result in normalization against changes in scale and contrast; finally, the condition c_{pq} positive real results in normalization against rotations.

2.9.4 Moments of the normalized image

Udagawa [47] first pointed out that one can obtain features invariant to affine transformations by using the moments of the normalized image, thus avoiding the need to actually transform the image to obtain a normalized version, and Reeves *et al.* [4] have performed some experiments using them. The trick is to use the parameters C (contrast) and a, b, c and d of the normalizing transformation to compute the moments of the normalized image as functions of the moments of the original image, where the normalizing transformation is given by $x' = ax + by$ and $y' = cx + dy$, plus the change in contrast by C. The central moments of a centred image are given by

$$\mu_{pq} = \int_{-\infty}^{+\infty} \int_{-\infty}^{+\infty} x^p y^q f(x,y)\, dx\, dy,$$

and the moments of the normalized image $f'(x', y')$ are given by

$$
\begin{aligned}
\eta_{pq} &= \int_{-\infty}^{\infty} \int_{-\infty}^{\infty} x'^p y'^q f'(x', y')\, dx'\, dy' \\
&= C \int_{-\infty}^{\infty} \int_{-\infty}^{\infty} (ax + by)^p (cx + dy)^q f(x,y)|J|\, dx\, dy.
\end{aligned}
$$

Hence we see that η_{pq} is a polynomial in the moments μ_{pq} of order up to $p+q$; examples are presented in chapter 4 when dealing with invariance to affine transformations, along with some experimental results.

2.10 Rotation invariants of discrete images

In this section, we will look at two aspects of rotation invariants based on discrete images. First, we will consider whether and under what circumstances one can form functions of a sampled image that are absolutely invariant under rotations of the continuous image; then, we will consider whether orthogonal moments remain orthogonal when the image is sampled.

2.10.1 Absolute invariants of discrete images

This section addresses the following two questions:

1. Given I_0, an $N \times N$ sampled version of a continuous image, and I_θ, the $N \times N$ sampled version of the same image rotated by an angle θ, what conditions must be fulfilled for us to be able to compute I_θ given only I_0 and θ?

2. Under what conditions can we compute functions of the I_θ that remain absolutely invariant over all values of $\theta \in [0, 2\pi]$?

We are used to thinking of a sampled version of a bandlimited image as containing all the information in the continuous image, as long as the sampling rate is high enough. However, this is misleading because a bandlimited image has infinite spatial extent and hence requires an infinite number of samples to fully describe it. In the case of an $N \times N$ discrete image, we only have N^2 points with which to fully describe the continuous image; hence, the latter must be completely described by some known set of N^2 functions for the first question to be satisfied. The most straightforward case is when the continuous image is exactly representable by a linear combination of N^2 functions. If we let $f(x, y)$ represent the continuous image and $v_{pq}(x, y)$, $1 \leq p, q \leq N$, represent the N^2 basis functions, then we want

$$f(x, y) = \sum_{p=1}^{N} \sum_{q=1}^{N} d_{pq} v_{pq}(x, y) \qquad (2.16)$$

for some set of coefficients $\{d_{pq}\}$. Let the sampled image be $g(k, l)$, then

$$g(k, l) = f(kx_0, ly_0), \qquad k, l = -\frac{N}{2}, -\frac{N}{2} + 1, \ldots, \frac{N}{2}, \qquad (2.17)$$

and N is assumed to be even for the sake of simplicity. Combining (2.16) and (2.17) gives

$$g(k, l) = \sum_{p=1}^{N} \sum_{q=1}^{N} d_{pq} v_{pq}(kx_0, ly_0), \qquad (2.18)$$

or $\qquad g_{kl} = V_{pqkl} d_{pq} \qquad$ in tensor notation.

Rather than using tensor notation, we can convert the matrix of values g_{kl} into an N^2-dimensional vector $\mathbf{g}$ with entries $g_i = g(k, l)$, $i = k + N/2 + 1 + N * (l + N/2)$. This allows us to write (2.18) as

$$\mathbf{g} = \mathbf{V}\mathbf{d}; \qquad \mathbf{V} \text{ is an } N^2 \times N^2 \text{ matrix.}$$

We can now see that the condition of question 1 will be satisfied if the continuous image I_0 is exactly represented by a linear combination of N^2 basis functions $v_{pq}(x, y)$ that result in a non-singular matrix $\mathbf{V}$. We can determine the coefficients using $\mathbf{d} = \mathbf{V}^{-1}\mathbf{g}$.

Now we are in a position to address the second question. Before doing this though, it should be noted that the previous result only assumes that I_0 can be exactly represented by the N^2 basis functions; no assumption is made about I_θ being represented by the $v_{pq}(x, y)$. In fact, if $\mathbf{V}$ is invertible, every I_θ, $\theta \in [0, 2\pi]$, will in general produce a different vector of coefficients $\mathbf{d}$ for each different value of θ, but combining these coefficients with the basis set as in (2.16) will generally result in a different continuous function from that of the rotated continuous image. If we are to compute invariants we would like I_θ to be exactly represented by the N^2 basis functions for all values of θ i.e. letting $f(r, \theta)$ be the image in polar coordinates and $\{v_{pq}(r, \theta)\}$ the basis in polar coordinates, we want the following to be satisfied: for all $\{d_{pq}\}$ and ϕ there exists $\{d'_{pq}\}$ such that

$$f(r, \theta + \phi) = \sum_p \sum_q d_{pq} v_{pq}(r, \theta + \phi) = \sum_p \sum_q d'_{pq} v_{pq}(r, \theta). \qquad (2.19)$$

If the above is satisfied we can hope to find how $\{d_{pq}\}$ is related to $\{d'_{pq}\}$ and hence functions of the d_{pq} that are invariant to rotations.

Equation (2.19) will hold if the set of functions $\{v_{pq}(r, \theta + \phi)\}$ spans the same space as $\{v_{pq}(r, \theta)\}$ for all ϕ. To make the notation easier, let $v_k(r, \theta) = v_{pq}(r, \theta)$, where $1 \le q \le Q$ and $k = p + Q * (q - 1)$. $\{v_k(r, \theta)\}$ spans the same space as $\{v_k(r, \theta + \phi)\}$ for all ϕ if

$$v_k(r, \theta + \phi) = \sum_{n=1}^{N^2} G_{kn}(\phi) v_n(r, \theta) \qquad \text{for all } \phi, \ k = 1, \ldots, N^2, \qquad (2.20)$$

where $G_{kn}(\phi)$ is some periodic function with period 2π. (2.20) can be written using vectors and matrices by defining $\mathbf{v}(r, \theta)^T = [v_1(r, \theta), \ldots, v_{N^2}(r, \theta)]$ and $\mathbf{G}(\phi)$ as the matrix with entry $G_{kn}(\phi)$ in row k and column n:

$$\mathbf{v}(r, \theta + \phi) = \mathbf{G}(\phi) \mathbf{v}(r, \theta), \qquad (2.21)$$

and $\mathbf{G}(\phi)$ is periodic with period 2π. (2.21) already tells us that $\mathbf{v}(r, \theta)$ must be a separable function: $v_k(r, \theta) = h_k(r) s_k(\theta)$, so (2.21) becomes

$$\mathbf{s}(\theta + \phi) = \mathbf{G}(\phi) \mathbf{s}(\theta). \qquad (2.22)$$

Considering the case $\phi = 0$, and using the fact that $\mathbf{s}(\theta) = \mathbf{G}^{-1}(\phi) \mathbf{s}(\theta + \phi)$, gives us

$$\begin{aligned} \mathbf{G}(0) &= \mathbf{I}, \quad \text{the } N^2 \times N^2 \text{ identity matrix;} \\ \mathbf{G}(-\phi) &= \mathbf{G}^{-1}(\phi). \end{aligned} \qquad (2.23)$$

An example of matrices $\mathbf{G}(\phi)$ that satisfy (2.23) are the rotation matrices that define a rotation by an amount ϕ about some axis in N^2 dimensions.

In order to find invariants given a matrix $\mathbf{G}(\phi)$, we need first to find the set of functions $\mathbf{s}(\theta)$ that satisfies (2.22) before proceeding to see how the coefficients $\{d_k\}$ in (2.19) change with ϕ. In general finding the form of $\mathbf{s}(\theta)$ is not easy; however, two special cases allow us to generate invariants easily. The first is when $\mathbf{G}(\phi)$ is diagonal, and hence $s_k(\theta + \phi) = G_{kk}(\phi) s_k(\theta)$. It is easy to show that in this case the only suitable function $G_{kk}(\phi)$ is $G_{kk}(\phi) = \exp(jl_k\phi)$, l_k an integer. Typically $l_k = k$. It is also straightforward to show that this value of $G_{kk}(\phi)$ results in $s_k(\theta) = \exp(jk\theta)$:

$$\mathbf{G}(\phi) \text{ diagonal} \Rightarrow G_{kk}(\phi) = e^{jl_k\phi} \Rightarrow s_k(\theta) = e^{jl_k\theta}.$$

Putting the above into (2.19) gives (remembering that $k = p + Q * (q - 1)$)

$$\sum_k d_k h_k(r) e^{jl_k(\theta + \phi)} = \sum_k d'_k h_k(r) e^{jl_k\theta}$$

from which we see that d'_k is related to d_k by

$$d'_k = d_k e^{jl_k\phi}.$$

This in turn tells us that the magnitude of the coefficients, $|d_k|$, is invariant to rotations of the continuous image.

Before describing the second special case, a quick word about using complex coefficients d_k. Since the image is real, it is described by N^2 values; however, a complex $\mathbf{d}$ has $2N^2$ parameters, so we expect a certain amount of redundancy among the $\{d_k\}$. The redundancy expresses itself in the fact that the l_k's form pairs (l_{k1}, l_{k2}) where $l_{k1} = -l_{k2}$. This means that only half the coefficients d_k are independent.

The second special case is when $\mathbf{G}(\phi)$ is of the form

$$\mathbf{G}(\phi) = \begin{bmatrix} & \vdots & \vdots & \vdots & \vdots & \vdots & \\ \cdots & 0 & 0 & 0 & 0 & 0 & \cdots \\ \cdots & \cos l\phi & -\sin l\phi & 0 & 0 & 0 & \cdots \\ \cdots & \sin l\phi & \cos l\phi & 0 & 0 & 0 & \cdots \\ \cdots & 0 & 0 & \cos(l+1)\phi & -\sin(l+1)\phi & 0 & \cdots \\ \cdots & 0 & 0 & \sin(l+1)\phi & \cos(l+1)\phi & 0 & \cdots \\ \cdots & 0 & 0 & 0 & 0 & \cos(l+2)\phi & \cdots \\ & \vdots & \vdots & \vdots & \vdots & \vdots & \end{bmatrix},$$

where l is an integer, which gives us (from (2.22))

$$\begin{aligned} s_k(\theta + \phi) &= \cos l\phi\, s_k(\theta) - \sin l\phi\, s_{k+1}(\theta); \\ s_{k+1}(\theta + \phi) &= \sin l\phi\, s_k(\theta) + \cos l\phi\, s_{k+1}(\theta). \end{aligned}$$

It is easy to show that these equations are satisfied by

$$s_k(\theta) = \cos l\theta, \qquad s_{k+1}(\theta) = \sin l\theta.$$

Invariants can be found by considering the coefficients of $v_1(r, \theta)$ and $v_2(r, \theta)$ with $l = 1$:

$$d_1 \cos(\theta + \phi) + d_2 \sin(\theta + \phi) = d'_1 \cos\theta + d'_2 \sin\theta.$$

Expanding $\cos(\theta + \phi)$ and $\sin(\theta + \phi)$ gives

$$d_1 [\cos\phi \cos\theta - \sin\phi \sin\theta] + d_2 [\sin\phi \cos\theta + \cos\phi \sin\theta] = d'_1 \cos\theta + d'_2 \sin\theta.$$

Comparing coefficients of $\cos\theta$ and $\sin\theta$ gives

$$\begin{aligned} d'_1 &= d_1 \cos\phi + d_2 \sin\phi; \\ d'_2 &= -d_1 \sin\phi + d_2 \cos\phi; \end{aligned}$$

from which we see that $d'^2_1 + d'^2_2 = d^2_1 + d^2_2$. This extends to the cases where $l \neq 1$, so $d^2_k + d^2_{k+1}$, k even, is invariant to image rotations.

Both special cases correspond to modeling the image with an angular Fourier series; in the second case we do not have the burden of using complex functions.

In conclusion we have seen in this section that the continuous image must be exactly representable by N^2 known basis functions to obtain invariants of the sampled version; we have also seen two equivalent cases where the invariants are easy to compute.

2.10.2   Orthogonal rotation invariants of discrete images

Below we will look at how the standard technique of replacing the integrals in the definition of moment invariants by summations results in the orthogonal invariants no longer being invariant, and we will see two ways of circumventing this to result in truly orthogonal features.

Consider the orthogonal polynomials of section 2.4.3, and let them be represented by $v_{pq}(x, y)$, $p, q = 1, 2, \ldots$. The moments z_{pq} of an image $f(x, y)$ are formed as follows:

$$z_{pq} = \int_{-\infty}^{+\infty} \int_{-\infty}^{+\infty} v_{pq}^*(x, y) f(x, y) \, dx \, dy, \tag{2.24}$$

where v^* is the complex-conjugate of v. If we replace the integrals by summations and adopt the notation of the previous section, we can see that (2.24) becomes $\mathbf{z} = \mathbf{V}^H \mathbf{g}$. In other words, $\mathbf{V}^H$ is used instead of $\mathbf{V}^{-1}$. The condition of orthogonality in the discrete case becomes $\mathbf{V}^H \mathbf{V} = \mathbf{I}$, where $\mathbf{I}$ is the $M \times M$ identity matrix, and M is the number of points in the image (in the previous section M was equal to N^2; if for example we use a square grid of points defined over the unit circle, $M \neq N^2$). It turns out that $\mathbf{V}^H \mathbf{V} \neq \mathbf{I}$; in fact, if one has P radial functions, appendix A shows that it is of the form

$$
\mathbf{V}^H \mathbf{V} =
\begin{bmatrix}
\mathbf{A} & 0 & 0 & 0 & \mathbf{B}_1 & 0 & 0 & 0 & \mathbf{B}_2 & 0 & \ldots \\
0 & \mathbf{A} & 0 & 0 & 0 & \mathbf{B}_1 & 0 & 0 & 0 & \mathbf{B}_2 & \ldots \\
0 & 0 & \mathbf{A} & 0 & 0 & 0 & \mathbf{B}_1 & 0 & 0 & 0 & \ldots \\
0 & 0 & 0 & \mathbf{A} & 0 & 0 & 0 & \mathbf{B}_1 & 0 & 0 & \ldots \\
\mathbf{B}_1 & 0 & 0 & 0 & \mathbf{A} & 0 & 0 & 0 & \mathbf{B}_1 & 0 & \ldots \\
\vdots & \vdots & \vdots & \vdots & \vdots & \vdots & \vdots & \vdots & \vdots & \vdots &
\end{bmatrix}
\qquad
\begin{array}{l}
\mathbf{A}, \mathbf{B}_i \text{ are} \\
P \times P \text{ matrices.}
\end{array}
$$

The orthogonal functions encountered earlier use polar separable functions of the form $h_p(r)e^{jq\theta}$. Let us assume we have P radial functions and Q angular ones such that $PQ = M$; the appendix also shows that $\mathbf{A}$ and $\mathbf{B}_i$ are $P \times P$ matrices, with elements in row p' and column p given by

$$
\begin{aligned}
(\mathbf{A})_{p'p} &= \sum_i r_i h_{p'}^*(r_i) h_p(r_i); \\
(\mathbf{B}_k)_{p'p} &= \sum_i r_i h_{p'}^*(r_i) h_p(r_i) e^{j4k\theta_i}.
\end{aligned}
$$

This tells us that the matrix $\mathbf{V}$ of the orthogonal functions with angular components $e^{jq\theta}$ are more nearly orthogonal than one might have supposed. Nevertheless, $\mathbf{V}^H \mathbf{V}$ is not equal to the identity matrix; even for images with reasonable resolution, such as those with a diameter of 64 pixels, the non-orthogonality is quite noticeable. Using pseudo-Zernike functions up to order 12 (i.e. $P = 13$), we find that $(\mathbf{A})_{PP} = 0.9186$, and $(\mathbf{A})_{P,P-1} = 0.1235$; ideally they should be 1 and 0 respectively. If we have a circular image with diameter 16 pixels, there are 180 pixels in the image; if we take $P = 5$ and $Q = 36$ we find that $\mathbf{A}$, $\mathbf{B}_1$ and $\mathbf{B}_2$ for the pseudo-Zernike basis are given to four decimal places by

$$\mathbf{A} = \begin{bmatrix} 1 & 0.0212 & 0.0275 & 0.0408 & 0.0435 \\ 0.0212 & 1.0166 & 0.0471 & 0.0472 & 0.0952 \\ 0.0275 & 0.0471 & 1.0342 & 0.1000 & 0.0577 \\ 0.0408 & 0.0472 & 0.1000 & 1.0428 & 0.2016 \\ 0.0435 & 0.0952 & 0.0577 & 0.2016 & 1.0267 \end{bmatrix} ;$$

$$\mathbf{B}_1 = \begin{bmatrix} -0.0108 & 0.0571 & -0.0435 & 0.1051 & 0.0002 \\ 0.0571 & -0.0883 & 0.1978 & -0.1163 & 0.2133 \\ -0.0435 & 0.1978 & -0.1792 & 0.3271 & -0.0652 \\ 0.1051 & -0.1163 & 0.3271 & -0.1380 & 0.3255 \\ 0.0002 & 0.2133 & -0.0652 & 0.3255 & 0.0842 \end{bmatrix} ;$$

$$\mathbf{B}_2 = \begin{bmatrix} 0.0094 & 0.0074 & 0.0538 & -0.0283 & 0.1265 \\ 0.0074 & 0.0611 & -0.0568 & 0.2088 & -0.1132 \\ 0.0538 & -0.0568 & 0.2248 & -0.1667 & 0.3275 \\ -0.0283 & 0.2088 & -0.1667 & 0.3512 & -0.0711 \\ 0.1265 & -0.1132 & 0.3275 & -0.0711 & 0.2145 \end{bmatrix} .$$

Obtaining truly orthogonal features

There are two ways of obtaining truly orthogonal features for discrete images; one is simply to orthogonalize the vectors obtained above, for example using the Gram-Schmidt procedure. Alternatively, one can be more cunning and use interpolation to transform the discrete image into a continuous one. If one uses the Zernike moments, which can be expressed as functions of the central moments [15], then the use of interpolation is particularly straightforward: the central moments of the interpolated image can be expressed as polynomials in the central moments of the discrete image computed using summations. Mardia & Hainsworth [50] suggest using a 'hat' interpolating function

$$\text{'Hat':} \quad \hat{h}(x,y) = \hat{g}(x)\hat{g}(y), \qquad \hat{g}(x) = \begin{cases} 1, & -\tfrac{1}{2} \le x \le \tfrac{1}{2} \\ 0, & \text{elsewhere,} \end{cases}$$

whereas the author has considered using a 'triangle' interpolating function, resulting in linear interpolation:

$$\text{'Triangle':} \quad h_t(x,y) = g_t(x)g_t(y), \qquad g_t(x) = \begin{cases} 1-x, & 0 \le x \le 1; \\ 1+x, & -1 \le x < 0; \\ 0 & \text{elsewhere.} \end{cases}$$

One can show that, if μ_{pq} are the central moments of the discrete image computed using summations in place of integrals in (2.2), then the central moments of the 'hat'-interpolated image, $\hat{\mu}_{pq}$, are given by [50]:

$$\hat{\mu}_{pq} = \frac{1}{(p+1)(q+1)} \sum_{r=0}^{\lfloor p/2 \rfloor} \sum_{s=0}^{\lfloor q/2 \rfloor} \binom{p+1}{2r+1} \binom{q+1}{2s+1} \left(\frac{1}{4}\right)^{r+s} \mu_{p-2r,\,q-2s},$$

and the central moments of the 'triangle'-interpolated image, η_{pq}, are given by:

$$\eta_{pq} = \frac{1}{(p+1)(p+2)(q+1)(q+2)} \sum_{r=0}^{\lfloor p/2 \rfloor} \sum_{s=0}^{\lfloor q/2 \rfloor} \binom{p+2}{2r+2} \binom{q+2}{2s+2} \left(\frac{1}{4}\right)^{r+s} \mu_{p-2r, q-2s},$$

where $\lfloor x \rfloor$ is the largest integer k such that $k \leq x$, and

$$\binom{p}{k} = \frac{p!}{k!\,(p-k)!}.$$

Experiments using these interpolated moments on binary images were performed using affine invariants, but there was little to choose between straightforward moments and interpolated ones. Note that, for binary images, Jiang & Bunke's [51] fast method is equivalent to computing moments of an interpolated image, and the above equations need not be used.

2.11   Correlation invariants for discrete images

We saw in section 2.4.1 that angular correlations for rotational invariance are not a natural extension of the continuous case because of the varying number of pixels at a given radius. One approach that sidesteps this problem is to sum over the angle variable and the radius variable, as with the radial-Mellin-circular-Fourier transform defined in section 2.6.1 but with the integrals replaced by summations. If one wants to use correlations, one can proceed by resampling the image using interpolation to estimate values between pixels or one can adopt the ingenious method introduced by Perantonis & Lisboa [52], which is summarized below.

The technique is best explained from a group-theoretical viewpoint; below, I have elected to present it in an essentially qualitiative manner; the reader seeking mathematical rigour is referred to Lenz's book [34].

As we will see, group theoretical methods lend themselves particularly well to classifiers based on correlations. The fundamental principle is to assign sets of points to equivalence classes and then to integrate over all members of the class. As a simple example, assume that we wish to obtain invariance to translations for an image with intensity $f(\mathbf{x})$ at a point $\mathbf{x} = [x\ y]^T$. The integrals below are over the range $[-\infty, +\infty]$.

Single points

Any single point can be translated to any other, so an invariant g can be obtained by integrating over the whole image:

$$g = \int f(\mathbf{x})\, d\mathbf{x}.$$

g is simply the first order correlation encountered in section 2.2.

Pairs of points

Any pair of points separated by a vector $\mathbf{v}$ can be translated to any other such pair of points. Hence pairs of points separated by $\mathbf{v}$ are equivalent under translation i.e. they form an equivalence class. Given any symmetric function $S(f_1, f_2)$, we obtain an invariant to translation by setting $f_1 \equiv f(\mathbf{x})$, $f_2 \equiv f(\mathbf{x} + \mathbf{v})$ and integrating over $\mathbf{x}$. For example, using correlations gives

$$g(\mathbf{v}) = \int f(\mathbf{x})\, f(\mathbf{x} + \mathbf{v})\, d\mathbf{x},$$

the second order correlation (see section 2.2). The same principle applies to triplets, quartets etc. of points.

Giles & Maxwell [53] were the first to consider using the above correlation invariants to obtain translation invariant classifiers for discrete images. As long as an input image is translated by an integer number of pixels, the object is on a zero intensity background and always fits entirely on the sampling lattice, the above results can be used for discrete images by changing the integrals to summations over pixels $\{x_i\}$.

$$g = \sum_i f(\mathbf{x}_i); \qquad g(\mathbf{v}_j) = \sum_i f(\mathbf{x}_i)\, f(\mathbf{x}_i + \mathbf{v}_j); \qquad \text{etc.}$$

2.11.1 Invariance to translations and rotations of discrete images

In section 2.4 we saw that one can obtain invariance to rotations of continuous images by using correlations in polar coordinates. Let us look at an alternative approach to obtaining rotation invariance for continuous images, before looking at how to do so for discrete images.

Using the argument of the previous section, it is clear that g will be invariant to translations and rotations. The equivalence class for a pair of points changes though: pairs of points separated by a vector $\mathbf{v}$ such that $|\mathbf{v}| = d$ are now equivalent to other such pairs, giving us the invariant

$$g_r(d) = \int_{|\mathbf{v}|=d} g(\mathbf{v})\, d\mathbf{v}$$

(Note that $g(\mathbf{v})$ has two parameters, whereas $g_r(d)$ has only one).

The above integral is difficult to turn into a summation for discrete images, because a point $\mathbf{x}$ will generally only have eight points $\mathbf{x} + \mathbf{v}$ on the sampling lattice such that $|\mathbf{v}| = d$, where $d^2 = a^2 + b^2$ for some integers a, b. Instead, we can adopt the technique introduced by Perantonis & Lisboa [52] and quantize d into intervals I_k, $k = 1, 2, \ldots$. Having done this, the above integral becomes the following summation:

$$\hat{g}_r(k) = \sum_{|\mathbf{v}| \in I_k} g(\mathbf{v}).$$

(Note that $g_r(d)$ is a continuous function and $\hat{g}_r(k)$ is discrete). In practice one would choose an upper limit on d to obtain a finite number of intervals. The above is even simpler if the image is binary: one assigns a bin b_k to each interval I_k and counts the number of times n_k that two non-zero image pixels appear with separation $|\mathbf{v}| \in I_k$ for each bin b_k, $k = 1, \ldots, N$.

2.11.2   Invariance to translation, rotation and scaling

Perantonis & Lisboa [52] used the above quantization method to obtain invariance to translations, rotations and changes in scale. Changes in scale force one to either scale the axes logarithmically (c.f. the Mellin transform) or to introduce a normalizing factor; Perantonis & Lisboa choose to do the latter. They show that one can obtain invariance to the above three transformations, up to a scale factor, by using the equivalence class of three points. The equivalence class they choose is the pair of angles (θ, ϕ) defined as the largest and second largest angles of the triangle defined by the three points (they also take into account the relative position of the angles so as to be able to discriminate between reflections).

Linear scaling of a binary image by a factor s scales the number of non-zero pixels in the image by approximately s^2, resulting in a factor $(s^2)^3 = s^6$ change in the number of triplets of non-zero pixels. Perantonis & Lisboa remove the scaling factor by forming a vector $\mathbf{n} = [n_1 \quad n_2 \cdots n_k]$ of the entries in each bin and normalizing $\mathbf{n}$ to have a standard Euclidean length. They compared the performance of their correlation invariants with that of invariants based on Zernike moments, using both a single- and a double-layer perceptron to classify the typed digits 0, 1,..., 9 on a 20×20 grid, and found that the correlation invariants performed better. In each case the two-layered perceptron with 40 hidden nodes gave best results, with a 91% success rate using moment invariants and a 97% success rate using correlations.

2.12   Performance and computational cost

2.12.1   Computational complexity

First we will look at the computational complexity of computing correlation invariants and then at that of computing moment invariants.

Correlations

Correlation invariants are ideally computed using the magnitude of the Fourier transform; as Casasent & Psaltis [28] demonstrate, the Fourier transform is best computed optically using the fact that it is proportional to the projection of an image through a lens onto a plane closer to the lens than the image's focal plane. Numerically computing the Fourier transform of an image with N^2 pixels requires $\propto N^4(\log_2 N)^2$ complex multiplications to compute N^2 features.

If one uses the discrete correlations of the previous section, the computation required is generally heavy. To compute the pth-order correlation of an input vector of length M requires the evaluation of $C_p^M = M!/p!\,(M-p)!$ p-tets of points; for an $N \times N$ image, $M = N^2$, so the computational complexity is $O(N^6)$ when using triplets of points. M is reduced somewhat when classifying binary images, since one need only consider the $M = N_I$ non-zero points. Nevertheless, the computational complexity is very high: Perantonis & Lisboa use $N^2 = 400$ pixels; if one assumes only half of them are non-zero, so $M = 200$, we still need to evaluate over 1.3×10^6 triplets. This is clearly feasible on modern computers, but generally one would like to classify images with much higher resolution than 20×20, in which case the computation required becomes prohibitive.

Algorithms	Number of Additions	Number of Multiplications
Straightforward	$10MN$	$20MN$
Direct	$10MN$	$18MN$
Delta	$(N+6)M$	$25M$
Recursive	$8MN + 15N$	0

Table 2.1: Computation Required to Compute 10 Image Moments.

One approach to limit the increase in computation as the resolution increases is to make use of the fact that all the information in binary images is stored in the boundary, by placing M equispaced points along the boundary and computing their correlations. With any luck, if M is large enough, the fact that one is not using a fixed starting point should not matter too much. This idea is discussed further in chapter 4, section 4.5.2, in order to obtain invariance to affine transformations.

Moments

Although in principle one could compute an infinite number of moments of a continuous image, in practice one uses discrete images and only lower order moments are used (see below). For discrete images the integrals in (2.2) are replaced by summations, resulting in the straightforward algorithm. Chen [54] reviews the computational complexity of four algorithms for computing moments: two standard ones, Hatamian's fast recursive algorithm [55] and Zakaria *et al.*'s [56] fast (delta) algorithm for binary images. For an $M \times N$ image the complexity of computing the ten moments $m_{00}, m_{01}, m_{10}, \cdots, m_{21}, m_{30}$ is given in table 2.1. Pan [57], however, shows that Chen's analysis of parallel implementations is flawed. If one limits oneself to binary images, one can speed up the computation by using Green's theorem [58]:

$$m_{pq} = \frac{1}{p+1} \int_C x^{p+1} y^q \, dy = -\frac{1}{q+1} \int_C x^p y^{q+1} \, dx. \qquad (2.25)$$

Li & Shen and Jiang & Bunke combine the recursive method with Green's theorem to obtain an algorithm that requires $O(N)$ operations for an $N \times N$ image. The two algorithms only differ in the fact that Li & Shen simply replace the integrals in (2.25) with summations, whereas Jiang & Bunke assume that neighbouring pixels on the boundary are joined by straight lines. As a result, Li & Shen's method is faster, requiring no multiplications and $O(N)$ additions; however, experiments performed by the author indicate that their technique performs significantly worse than the straightforward method; Jiang & Bunke's method is likely to perform better because they compute a true line integral, rather than approximating it as Li & Shen do.

2.12.2 Performance

In the low noise case, moment invariants based on the central moments are handicapped by the fact that the basis $\{x^p y^q\}$ is not orthogonal, which introduces a large amount of correlation between different moments, and leads to a significant amount of redundant information [36]. This means that an infeasibly large number of moments would have to be computed to discriminate between images that only differ in their high frequency characteristics, from which we see that moments are best suited to classifying objects whose energy is concentrated in the lower frequencies. The binary shape of an object, like an aircraft's silhouette, is a prime example of the type of images to which moments are well suited. The only moment invariants that do not suffer from information redundancy are the orthogonal rotation invariants.

Correlations will generally perform equally well for images with predominantly low frequencies or predominantly high frequencies. The rotation invariants are however sensitive to the resolution of discrete images — as the resolution decreases, so the invariants will perform significantly less well. This is investigated in the next section.

Additive noise affects high order moments more than low order moments; Abu-Mostafa & Psaltis [36] show that the signal to noise ratio (SNR) of the complex moments c_{pq} asymptotically decreases as $1/\sqrt{p+q}$, and the plots of Teh & Chin [15] show the SNR decreasing with increasing moment order. However, as noted in chapter 1, in many cases the images one is dealing with have very low noise levels, in which case the above results are less important; often, a more significant cause of error is the distortions due to the use of discrete images, an effect which is investigated for moment-based rotation invariants in the next section.

Image correlations, not to be confused with ensemble-averaged correlations taken over a sequence of images, are also affected by noise; the higher the order of the correlation, the more they will be affected. In contrast to moments, each correlation (bar $g(\mathbf{0})$) will be equally affected by independent white noise.

2.13   Experimental results

This section describes experiments that were performed on rotationally invariant moment features; the goals were to ascertain the sensitivity of the invariants to the distortions caused by the discrete nature of the images and to determine how well the features discriminate between different shapes. Two sets of experiments were performed, one on the two binary shapes 'c' and 'τ' displayed in figures 2.7 and 2.8 and one on two rectangles of differing size and intensity.

2.13.1   Discriminating between two binary shapes

The two binary shapes 'c' and 'τ' were used to test the ability of the various moment-based features to discriminate between rotated versions of each letter. The shapes were sampled using a rectangular grid and are described in appendix D; eight horizontal or vertical pixels correspond to the width of the letter 'c'. Since the images are sampled coarsely, one can generate all possible sampled versions of the rotated shapes — these are depicted in Figs. 2.7 and 2.8, where we see that there are more distinct versions of the letter 'c' than the letter 'τ'. It should also be pointed out that the two shapes have identical radial distributions: if we let $f_c(r, \theta)$ be the image 'c' and $f_\tau(r, \theta)$ be the image 'τ', then

$$h_c(r) = \int_0^{2\pi} f_c(r, \theta)\, d\theta = h_\tau(r) = \int_0^{2\pi} f_\tau(r, \theta)\, d\theta.$$

Figure 2.7: The 16×16 versions of 'c' sampled over all orientations.

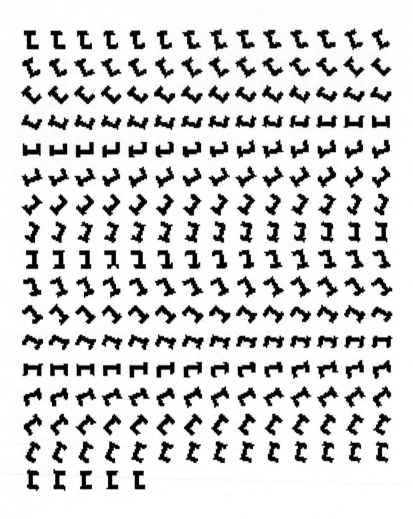

Figure 2.8: The 16×16 versions of 'τ' sampled over all orientations.

q:	7	1	1	1	3
s:	0	0	1	sin 1	sin 1
Mean 'c':	9.8062	8.7549	29.4498	1.2001	0.6828
Mean 'τ':	1.4741	16.8669	25.5916	0.9038	0.3332
Min. 'c':	7.5897	5.5078	27.7393	1.1385	0.5160
Max. 'c':	12.3402	11.1323	32.3412	1.3139	0.8585
Min. 'τ':	0.0390	14.4995	23.8047	0.8780	0.1992
Max. 'τ':	2.8287	18.3522	26.5407	0.9246	0.4316
Norm. Separation:	0.2509	0.1775	0.0627	0.2979	0.1175

Table 2.2: Features that discriminate between 'c' and 'τ'.

On the left are the features of the pseude-Zernike basis $h_s(r)e^{jq\theta}$, s being the order of the polynomial; on the right are those of the RWF basis $\{\sin k2\pi re^{jq\theta}, \cos k2\pi re^{jq\theta}\}$; $s = \sin 1$ means the radial function was $\sin 1.2\pi r$.

Mean 'c':	0.0272	0.0238	0.0117
Mean 'τ':	0.0043	0.0463	0.0201
Min. 'c':	0.0210	0.0151	0.0065
Max. 'c':	0.0336	0.0305	0.0180
Min. 'τ':	0.0004	0.0398	0.0159
Max. 'τ':	0.0072	0.0505	0.0239
Norm. Separation:	0.2617	0.1761	-0.0402

Table 2.3: The best three features using the pseudo-inverse $\mathbf{V}^{\#}$.

This means that one cannot discriminate between the two shapes using radial moments alone.

Each grid has 180 points, so using $\mathbf{V}^H$ gives us 180 features. For each of these features, the mean, maximum, minimum and standard deviation for each shape was computed, as was the minimum separation between the two shapes. For example, if a certain feature takes on values between 1 and 3 for the different orientations of the shape 'c' and values between 8 and 15 for the shape 'τ', the minimum separation is $8 - 3 = 5$. The minimum separation is then divided by $|\mathbf{v}|$, where $\mathbf{v}$ is the vector onto which the sampled image $\mathbf{g}$ is projected to get the feature d: $d = \mathbf{v}^T\mathbf{g}$.

Table 2.2 shows those features from the 180 of $\mathbf{V}^H$ that provided discrimination, and compares the pseudo-Zernike moments with the real weighted Fourier (RWF) moments. The angular function in both cases is $e^{jq\theta}$ — only features with $q \geq 0$ are displayed. Each such feature has an identical twin with $q' = -q$. The analysis above tells us tl.at the features are exactly orthogonal.

Table 2.3 shows the best three features using $\mathbf{V}^{\#}$; 24 columns of $\mathbf{V}$ were chosen: those for which $q = -12, \ldots, 11$ and $s = 0$ (the order of the pseudo-Zernike polynomials). As expected from table 2.2, only two of these features provide discrimination — the third feature in table 2.2 has $s = 1$. We see that the best feature performs slightly better than the best one using $\mathbf{V}^H$, but not as well as the best RWF feature.

| q: | 0 | 0 | 12 | 12 | 0 | 16 | 12 | 4 |
s:	1	2	2	1	sin 1	0	sin 1	sin 1
Mean 1:	83.35	36.36	39.28	40.81	0.1607	1.8044	1.5713	0.3167
Mean 2:	257.87	158.06	154.72	161.86	9.5832	6.8521	6.1761	4.5576
Min. 1:	77.09	32.54	33.29	34.62	0.0236	1.5902	1.3815	0.0130
Max. 1:	87.89	42.88	42.15	47.03	0.4133	2.0407	1.7917	0.5876
Min. 2:	237.69	148.33	146.55	140.98	8.9387	6.3328	5.4706	4.2688
Max. 2:	287.98	163.68	159.02	179.05	10.4235	7.3608	6.8357	4.7950
Sep.:	7.8307	5.4649	5.4110	4.9108	11.8755	5.5714	5.1246	4.7784

Table 2.4: The best four pseudo-Zernike and RWF features to discriminate between the two rectangles.

The larger rectangle is shape 1. The RWF features provide a larger normal-ized separation. The bottom row shows the normalized separation between the shapes. See table 6.1 for an explanation of rows 'q' and 's'.

2.13.2   Discriminating between two rectangles of differing size and intensity

When classifying different binary shapes the primary distinguishing feature will be the shape's area. If two shapes have the same area, we have already seen that their radial distributions can be identical; it is fairly easy to see that their angular distributions will only be the same if the shapes themselves are identical. The angular distribution can be the same if we use non-binary images; the results presented below are of experiments performed on a very simple example of two shapes with identical angular distributions: they are two rectangles, one twice the size of the other (and hence four times the area), but with a quarter the intensity of the smaller one.

Table 2.4 shows the best four features using $\mathbf{V}^H$ for the pseudo-Zernike basis and the RWF basis; in both cases many more features also provided discrimination between the shapes. Again we see that the RWF features perform significantly better. Note that this time the features are correlated with one another, so one ought to orthogonalise the vectors to remove the correlation.

2.14   Conclusions

In this chapter we have seen how one can either use image correlations or image moments to achieve invariance to image translation, rotation, changes of scale and changes of contrast. We have also looked at obtaining rotation invariants for discrete images, and we have seen that features that are orthogonal for continuous images are no longer orthogonal for discrete images, but that orthogonal features can be formed. The orthogonal Zernike and pseudo-Zernike moments were generalized, and an alternative orthogonal basis, the Real Weighted Fourier basis, was introduced and found to outperform the Zernike-type moments. We have also seen how to normalize a grey-scale image against changes in contrast and/or brightness; the experiment that will be described in chapter 4 (section 4.5.3) shows that it is in practice very important to take changes in both contrast and brightness into account, and good results are achieved using an image normalized against these changes.

Perantonis & Lisboa's [52] technique for using correlations of discrete images was discussed; according to their results, they outperform pseudo-Zernike moments when classifying 20×20 images. However, for an $N \times N$ image their computational complexity is $O(N^6)$, so it is not feasible for images with higher resolution. A refinement of their technique, with much lower computational complexity and invariance to affine transformations is discussed in chapter 4, section 4.4.3.

When using moments one can choose between using moment invariants or using invariants based on normalization; the major disadvantage of normalization is that it is more difficult when classifying objects with rotational symmetries. However, if combined with template matching it can provide good classification in the presence of additive noise [36].

Of the principles discussed in this chapter, those that apply to the affine invariants of chapter 4 are the moments and correlations defined in section 2.2, the complex moments (2.4.2), contrast invariance (2.8), normalization (2.9), discrete correlations (2.11) and the comments on the performance and computational cost (2.12). Before going on to discuss invariance to affine transformations however, we will need to review the theory of algebraic invariants, which is the subject of the next chapter.

Chapter 3

Algebraic and Projective Invariants

3.1   Introduction

The theory of algebraic invariants lies at the heart of features invariant to affine image transformations and projective image transformations. Research into algebraic invariants appears to have been initiated by Boole in 1841 [59], and was rapidly taken up by Cayley, Sylvester and Salmon in Britain and Ireland and by Clebsch, Gordan and others on the continent (references can be found below). Many invariants were explicitly listed, in particular by Salmon [60, 61]. Hilbert solved the main outstanding issues in the 1890s [62, 63], after which interest in the subject waned. These nineteenth-century 'historical' techniques have largely been superceded by the tensor-based method as discussed by Gurevich [64]. Dickson's book [65] provides a reasonably concise introduction to the historical techniques; further reviews can be found in references [59, 66].

Since the literature on computer vision does not contain an introduction to the theory and use of algebraic invariants, and the literature on the subject is fairly inaccessible, this chapter is devoted to filling the gap. The first section defines algebraic invariants in both two and three dimensions and shows how their existence can be interpreted both algebraically and geometrically, the latter case showing how algebraic invariants lead to projective invariants. The next section discusses the number of independent invariants that exist, and also the number of so-called irreducible invariants, and the final section lists a large number of algebraic and projective invariants, both for 2D and 3D; most of the 2D invariants are culled from Salmon's book [61], but a few new ones were generated by the author using both the historical and the modern methods. Appendix B contains a discussion of both the historic and the tensor based methods for generating the invariants.

The 2D invariants will be used in chapter 4 to obtain moment features invariant to affine transformations, while the 3D invariants will be used in chapter 5 to obtain invariance to planar projective transformations.

3.2 Introduction to algebraic and projective invariants

The first part of this section explains what algebraic invariants are and gives some simple examples; it also explains how the existence of invariants can be interpreted both algebraically and geometrically. The second part shows how functions invariant to projective transformations can be derived directly from algebraic invariants.

3.2.1 Algebraic invariants

Invariants of binary forms

Algebraic invariants arise from the study of homogeneous polynomials in a number of variables x, y, z etc. These polynomials are called *forms*; we will first look at binary forms, which have two variables x and y, before looking briefly at ternary forms, which have the three variables x, y, and z. A binary form of order p is defined as

$$f_p(x,y) = a_{p,0}x^p + \binom{p}{1}a_{p-1,1}x^{p-1}y + \binom{p}{2}a_{p-2,2}x^{p-2}y^2 + \ldots + a_{0,p}y^p.$$

$$\text{where} \quad \binom{p}{k} = \frac{p!}{k!\,(p-k)!}.$$

A simple example is the binary quadratic form $f_2(x,y)$ which, if we set $a = a_{2,0}$, $b = a_{1,1}$ and $c = a_{0,2}$, is given by

$$f_2(x,y) = ax^2 + 2bxy + cy^2. \tag{3.1}$$

How are the coefficients $\{a_{i,p-i}\}$ affected by a linear coordinate transformation given by

$$\begin{aligned} x &= \alpha x' + \beta y'; \\ y &= \gamma x' + \delta y'; \end{aligned} \qquad \Delta = \alpha\delta - \beta\gamma \neq 0?$$

If we substitute these expressions for x and y into equation (3.1) we get

$$\begin{aligned} f_2(x,y) &= a(\alpha x' + \beta y')^2 + 2b(\alpha x' + \beta y')(\gamma x' + \delta y') + c(\gamma x' + \delta y')^2 \\ &= a'x'^2 + 2b'x'y' + c'y'^2 \end{aligned}$$

where a', b' and c' are defined as

$$a' = a\alpha^2 + 2b\alpha\gamma + c\gamma^2, \qquad c' = a\beta^2 + 2b\beta\delta + c\delta^2,$$

$$b' = a\alpha\beta + b(\alpha\delta + \beta\gamma) + c\gamma\delta. \tag{3.2}$$

(Note that these equations are linear in the coefficients a, b, c.)

In the nineteenth century mathematicians asked themselves the following question: are there functions of the coefficients $\{a_{i,p-i}\}$ that remain unaffected by linear coordinate transformations? The answer is in the affirmative, and the resulting functions are the so-called algebraic invariants. A function $I(a_{p,0}, \ldots, a_{0,p})$ is called an invariant if

$$I(a'_{p,0}, \ldots, a'_{0,p}) = \Delta^g I(a_{p,0}, \ldots, a_{0,p}). \tag{3.3}$$

If the *weight* $g = 0$ then I is an *absolute* invariant; otherwise it is called a *relative* invariant. Given two relative invariants one can always form an absolute invariant by dividing suitable powers of the relative invariants to remove the Δ^g terms. A simple relative invariant is Q, that of the binary quadratic form $f_2(x, y)$; one can easily show that

$$Q' = a'c' - b'^2 = \Delta^2(ac - b^2) = \Delta^2 Q. \tag{3.4}$$

In other words, $Q = I(a, b, c) = ac - b^2$ is an invariant of weight 2. Looking at equation (3.2) we see that the binary quadratic gives us three linear equations relating new coefficients to old; in general $f_p(x, y)$ will result in $p + 1$ linear equations. A linear transformation requires four parameters to define it, so one expects $p - 3$ of the equations in the coefficients to be dependent on one another, allowing us to find $p - 3$ absolute invariants $(p > 3)$ [61].

A simple example of an absolute invariant is that of the binary quartic

$$f_4(x, y) = ax^4 + 4bx^3y + 6cx^2y^2 + 4dxy^3 + ey^4$$

which has two relative invariants [61]:

$$\begin{array}{llr} S & = & ae - 4bd + 3c^2, \qquad\qquad\qquad \text{weight} \quad g = 4; \\ T & = & ace + 2bcd - ad^2 - eb^2 - c^3, \qquad\qquad\qquad g = 6. \end{array} \tag{3.5}$$

Writing $\hat{I} = \Delta^g I$ as shorthand for equation (3.3) allows us to show that S^3/T^2 is an absolute invariant:

$$\frac{S^3}{T^2} = \frac{(\Delta^4\hat{S})^3}{(\Delta^6\hat{T})^2} = \frac{\Delta^{12}\hat{S}^3}{\Delta^{12}\hat{T}^2} = \frac{\hat{S}^3}{\hat{T}^2}.$$

So far we have only discussed invariants of a single binary form; invariants also exist for a system of binary forms, where the invariants are now polynomials in the coefficients of all the forms. The simplest example is the invariant of the two linear forms

$$l = ax + by, \qquad L = Ax + By.$$

Using the same notation as above one can easily show that

$$a'B' - b'A' = \Delta(aB - bA).$$

Another simple example is the joint invariant between the linear form $l = ax + by$ and the quadratic form $Ax^2 + 2Bxy + Cy^2$, given by

$$J_1 = a^2C - 2abB + b^2A, \qquad \text{weight } g = 2. \tag{3.6}$$

Further examples are presented in section 3.4.

Algebraic properties of invariants

What do algebraic invariants tell us about their forms? A relative invariant is zero when the form has a particular structure which is invariant under linear transformations. For example, the invariant of the binary form, $ac - b^2$, is zero if the form is a perfect square:

$$ac - b^2 = 0 \quad \Rightarrow \quad f_2(x, y) = (\sqrt{a}x + \sqrt{c}y)^2.$$

Replacing x by $\alpha x' + \beta y'$ and y by $\gamma x' + \delta y'$ will clearly result in a perfect square in x' and y': $f_2'(x', y') = (\sqrt{a'}x' + \sqrt{c'}y')^2$, which implies that $a'c' - b'^2 = 0$.

Invariants in terms of the roots

So far we have only considered algebraic invariants in terms of the coefficients $\{a_{i,p-i}\}$ of a form; one can also consider them in terms of the form's p roots, which is directly relevant to forming projective invariants. If we let the roots be $\{t_i\}$, $i = 1, \ldots, p$, then we can write $f_p(x, y)$ as

$$f_p(x, y) = a_{p,0}(x - t_1 y)(x - t_2 y) \cdots (x - t_p y). \tag{3.7}$$

As an example,

$$f_2(x, y) = ax^2 + 2bxy + cy^2 = a(x - t_1 y)(x - t_2 y)$$

$$\text{and} \qquad ac - b^2 = -\frac{a^2}{4}(t_1 - t_2)^2.$$

Similarly, if we take the two linear forms

$$l_1 = a_1(x - t_1 y), \qquad l_2 = a_2(x - t_2 y)$$

then the invariant is

$$aB - bA = a_1 a_2 (t_2 - t_1). \tag{3.8}$$

In both cases we can interpret the invariant being zero as meaning that the two roots coincide. Dickson [65] shows that all invariants of a single binary form are functions of the differences of the roots, and that they are symmetric in the roots i.e. exchanging two roots t_i and t_j will not alter the invariant. In fact, his theorem on p.55 states that

> **Theorem** (Dickson [65])
> Any invariant of order p and weight g of the binary form $a_{p,0} x^p + \ldots$ equals the product of $a_{p,0}$ by a sum of products of constants and certain differences of the roots, such that each root occurs exactly p times in every product; moreover, the sum equals a homogeneous symmetric function of the roots of total order g. Conversely, the product of any such sum by $a_{p,0}$ equals a rational integral invariant.

This will be used later to show that projective invariants are equivalent to algebraic invariants.

Invariants of ternary forms

Before looking at projective invariants, we need to take a quick look at the invariants of ternary forms. A ternary (3-D) form $f_p(x, y, z)$ is written as

$$f_p(x, y, z) = a_{p,0,0} x^p + a_{0,p,0} y^p + a_{0,0,p} z^p + \ldots + \binom{p}{k}\binom{k}{l} a_{p-k,k-l,l} x^{p-k} y^{k-l} z^l + \ldots \tag{3.9}$$

A pth order form has $\frac{1}{2}p(p + 3) + 1$ coefficients — hence a cubic ternary form has 10 coefficients and a quartic ternary form has 15 coefficients. A linear transformation in 3-D has 9 parameters, so a cubic will have one absolute invariant and a quartic five; examples of invariants are listed in section 3.4.2. As with the binary forms, the invariants are most easily generated using tensors [64, 67, 68].

3.2.2 Projective invariants from algebraic invariants

We saw in the section in the first chapter on viewing transformations that the projection of points on a line is governed by a 2-D linear transformation in homogeneous coordinates. We will see below that projective invariants can be obtained simply from algebraic invariants by considering the roots of an algebraic form as defining points on a line [69]. Just as the roots of a 1-D polynomial define points on a line, so the roots of a 2-D polynomial define curves on a plane; we will see below that this implies that polynomial curves of a certain order are projected to polynomial curves of the same order under planar projection.

If we define $t = x/y$, we see that the roots of $f_p(x, y)$ are the same as those of $f_p(x, y)/y^p$ (3.7):

$$g_p(t) = \frac{f_p(x, y)}{y^p} = a_{p,0}(t - t_1)(t - t_2) \cdots (t - t_p).$$

Consider the difference between two roots $t_2 - t_1$ under the 1-D projective transformation

$$t = \frac{\alpha t' + \beta}{\gamma t' + \delta}, \qquad \Delta = \alpha\delta - \beta\gamma :$$

$$t_2 - t_1 = \frac{\Delta(t_2' - t_1')}{(\gamma t_1' + \delta)(\gamma t_2' + \delta)}. \tag{3.10}$$

This shows that merely cancelling powers of Δ in numerator and denominator will not give a projective invariant — one needs to ensure that each root appears an equal number of times in each term in the numerator and the denominator. Fortunately, Dickson's above theorem guarantees that this will be the case for absolute invariants of a single binary form. Taking the quartic form, the simplest two invariants as a function of the roots are

$$
\begin{aligned}
S &= a_0^2 \left[(t_1 - t_2)^2(t_3 - t_4)^2 + (t_1 - t_3)^2(t_2 - t_4)^2 + (t_1 - t_4)^2(t_2 - t_3)^2 \right]; \\
M &= a_0^6 \left[(t_1 - t_2)(t_1 - t_3)(t_1 - t_4)(t_2 - t_3)(t_2 - t_4)(t_3 - t_4) \right]^2;
\end{aligned}
\tag{3.11}
$$

resulting in the projective invariant S^3/M. This invariant does not depend on the ordering of the four roots since it is symmetric in them; although this would be important if the polynomials were complex, it is not necessary for real polynomials since the order of points on a line is preserved under projection (except for a possible reversal). The cross-ratio ρ of four points on a line is the fundamental projective invariant and its value is unchanged by reversing the order of the points. It is defined by

$$\rho = \frac{(t_3 - t_1)(t_4 - t_2)}{(t_4 - t_1)(t_3 - t_2)},$$

and is depicted in figure 3.1. (N.B. Six distinct values of ρ can be obtained by permuting the labelling of the four points [65].) We can interpret ρ as being the joint invariant of the four lines $l_i = a_i(x - t_i y) = 0$ (see equation (3.8)), where each root appears equally often in numerator and denominator as required to cancel terms in (3.10).

We can extend the above approach to deal with planar (2-D) projection by considering the roots of $f_p(x, y, z)$ (3.9). By defining $X = x/z$ and $Y = y/z$ we find

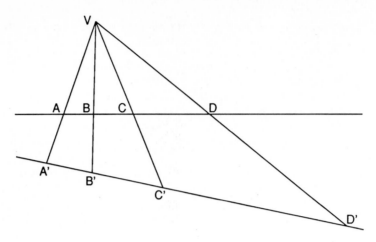

Figure 3.1: The cross-ratio.

The ratio AC/BC : AD/BD is an absolute invariant to projective transformations.

that

$$g_p(X, Y) = \frac{f_p(x, y, z)}{z^p} = a_{p,0,0}X^p + a_{0,p,0}Y^p + a_{0,0,p} + \cdots$$

i.e. $g_p(X, Y)$ is a pth order polynomial in X and Y, and its roots are pth order polynomial plane curves. Just as the roots of the binary form can be interpreted as points on a line, which are projected to a different set of points on a line, so polynomial plane curves are projected to different polynomial plane curves, but of the same order. The two examples of polynomial plane curves that occur most often in vision are lines and conics.

The algebraic invariants of four lines in homogeneous coordinates give rise to the projective invariants of four collinear points; similarly, the algebraic invariants of five planes in homogeneous coordinates give rise to the projective invariants of five coplanar lines. If the five coplanar lines are defined as $a_i x + b_i y + c_i = 0$, then one of the invariants is given by [9]:

$$(I_{431}I_{521})/(I_{421}I_{531}), \tag{3.12}$$

where I_{ijk} is the determinant of the matrix $[\ l_i\ \ l_j\ \ l_k\ ]$, with $l_i^T = [\ a_i\ \ b_i\ \ c_i\ ]$. Note that each root (line) appears the same number of times in numerator and denominator. Further invariants are listed in section 3.4.

3.3   The number of independent and irreducible invariants

Any function of an invariant is clearly also an invariant; an invariant is said to be independent if it cannot be expressed as a function of other invariants. Researchers in the nineteenth century looked for independent invariants and also for invariants that cannot be expressed as a polynomial in other invariants; Salmon [61] calls these invariants 'irreducible' to distinguish them from independent ones. For example, if we have three invariants P, Q, R and a fourth invariant M related to the others by $M = P^2+3QR$, M is not an irreducible invariant; however, if $M^2 = P^2+3QR$ then M is irreducible. Total alternation of high order products of moment tensors (discussed in appendix B) will lead to invariants that are simple powers and products of lower order invariants, as well as those that are not (the irreducible invariants); knowing how many irreducible invariants exist for a given set of image moments makes finding them using the tensor technique easier, since it tells one explicitly the order of the tensor products on which total alternation should be performed.

Cayley started the search for irreducible invariants, and proved that there are a finite number of them for cubic and quartic binary forms; however, he was unable to prove that this was the case for higher order binary forms. In 1868 Gordon proved by construction, using the symbolic method introduced Aronhold and Clebsch that, for a binary form of any order, there are a finite number of irreducible invariants. The question now was whether this also applied to forms in any number of variables. This question was finally answered in the affirmative by Hilbert in two seminal papers that appeared in 1890 and 1893 respectively [62, 63]. Thereafter mathematicians concentrated on building up group theory, and algebraic invariants became largely forgotten.

The irreducible invariants, although not independent of one another, are nevertheless useful additional features for two reasons: first, and most importantly, they contain sign information that cannot be obtained from the other invariants, and second, they are nonlinear functions of the other invariants and hence can enhance the ability of a classifier using them [70].

A binary form of order p has $p + 1$ coefficients; since a general linear transformation has four parameters, the binary form of order p will have $p - 3$ independent absolute invariants [61], and hence $p - 2$ independent relative invariants. What can we say about the number of irreducible invariants? As we saw above, it has been proven that there are a finite number of them. It is also possible to calculate the number of invariants a given form or system of forms will have of a given order in the coefficients, based on a method introduced by Cayley [71] and formally proven by Sylvester [72], which is summarized in appendix B. Using their technique, which is easily implemented on a computer, the number and order of the irreducible invariants of the system of binary, cubic and quartic forms have been computed and are listed in table 3.1. There are a total of 65 irreducible invariants, of which 38 are skew.

3.4   Examples of invariants

The first section below deals with invariants of binary forms, and twenty invariants, five of which are skew, are listed explicitly. The second section deals with invariants of ternary forms, and the invariants of points, lines, conics and of the ternary quartic are listed.

3.4.1   Invariants of binary forms

The invariants G_1, G_2 and K presented below have not been listed before; the remaining invariants are listed by Salmon [61]. G_1 and K were generated using the historic techniques, and G_2 using tensors on a computer. The tensor program was tested by

k	Invariants of the System of Quadratic, Cubic and Quartic Forms					
2	Q (2,0,0):1	S (0,0,2):1				
3	L (2,0,1):1	I (1,2,0):1	T (0,0,3):1			
4	N (2,0,2):1	G (1,2,1):2	P (0,4,0):1			
5	R (3,2,0):1	(2,2,1):2	(1,2,2):3	K (0,4,1):1	Z (0,2,3):1	
6	(3,2,1):1	H (3,0,3):1	(2,2,2):3	(1,4,1):2	(1,2,3):2	(0,4,2):2
7	(4,2,1):1	M (3,4,0):1	(3,2,2):1	(2,4,1):1	(2,2,3):3	(1,4,2):4
	(1,2,4):1	(0,4,3):3				
8	(4,2,2):1	(3,4,1):1	(2,2,4):1	(1,4,3):5	(0,6,2):1	(0,4,4):2
9	(1,4,4):3	(0,6,3):3	(0,4,5):1			
10	(1,4,5):1	(0,6,4):2				
11	(0,6,5):1					

Table 3.1: Irreducible invariants of the system of binary, cubic and quartic forms.

If there are N invariants of order θ_2 in the coefficients of the quadratic, order θ_3 in those of the cubic and order θ_4 in those of the quartic, this is represented as $(\theta_2, \theta_3, \theta_4)$: N. The total order k of the invariant is given in the leftmost column. Those entries in the table preceded by a capital letter are listed in full in the text.

generating the invariants L and H. It is worth noting (§143, [61]) that a binary form of order n whose coefficients have order θ in an invariant contributes an amount $\frac{1}{2}n\theta$ to the weight of the invariant. For example, take the invariant I of the cubic and quadratic forms below: for the cubic, $n = 3$, $\theta = 2$ and for the quadratic $n = 2$, $\theta = 1$. Hence the weight is $\frac{1}{2}.3.2 + \frac{1}{2}.2.1 = 4$ (this also follows easily from the tensor theory presented in appendix B).

In what follows, a^i represents the tensor of coefficients of the linear form, b^{ij} that of the quadratic form, c^{ijk} that of the cubic form and d^{ijkl} that of the quartic form (see appendix B).

The system of linear and quadratic forms

The joint invariant of the linear form $l = ax + by$ and the quadratic form $Ax^2 + 2Bxy + Cy^2$ is given by

$$J_1 = a^2 C - 2abB + b^2 A, \qquad \text{weight } g = 2,$$

and is equal to $a^i a^j b^{kl} \epsilon_{ik} \epsilon_{jl}$.

The system of linear and cubic forms

The joint invariant of the linear form $l = ax + by$ and the cubic form $\alpha x^3 + 3\beta x^2 y + \gamma xy^2 + \delta y^3$ is given by

$$J_2 = a^3 \delta - 3a^2 b\gamma + 3ab^2 \beta - b^3 \alpha \qquad g = 3,$$

and is equal to $a^i a^j a^k b^{lmn} \epsilon_{il} \epsilon_{jm} \epsilon_{kn}$.

The system of linear, quadratic and cubic forms

The joint invariant of the linear and cubic forms in the previous example with the quadratic form $Ax^2 + 2Bxy + Cy^2$ is given by

$$J_3 = a(A\delta - 2B\gamma + C\beta) - b(A\gamma + 2B\beta - C\alpha) \qquad g = 3,$$

and is equal to $a^i b^{jk} c^{lmn} \epsilon_{il} \epsilon_{jm} \epsilon_{kn}$.

In the following A, B, C will represent the coefficients of the binary form $Ax^2 + 2Bxy + Cy^2$; α, β, γ, δ those of the cubic form $\alpha x^3 + 3\beta x^2 y + \ldots + \delta y^3$ and $a, b, \ldots, e$ those of the quartic form $ax^4 + 4bx^3 y + \ldots + ey^4$ (the quintic and sextic forms are the only exceptions).

The quadratic form

Invariant: $Q = AC - B^2$ with weight $g = 2$.

The cubic form

Invariant: $P = (\alpha\delta - \beta\gamma)^2 - 4(\alpha\gamma - \beta^2)(\beta\delta - \gamma^2)$, $\qquad g = 6$.

The system of cubic and quadratic forms

Invariants additional to those of the cubic, P, and quadratic, Q, alone are I, R and M, of weight $g = 4$, 6 and 9 respectively:

$$I = A(\beta\delta - \gamma^2) - B(\alpha\delta - \beta\gamma) + C(\alpha\gamma - \beta^2),$$

$$\begin{aligned} R = {} & \alpha^2 C^3 - 6\alpha\beta BC^2 + 6\alpha\gamma C(2B^2 - AC) + \alpha\delta(6ABC - 8B^3) + 9\beta^2 AC^2 \\ & - 18\beta\gamma ABC + 6\beta\delta A(2B^2 - AC) + 9\gamma^2 A^2 C - 6\gamma\delta A^2 B + \delta^2 A^3, \end{aligned}$$

$$\begin{aligned} M = {} & A^3(3\beta\gamma\delta^2 - \alpha\delta^3 - 2\gamma^3\delta) + 6A^2 B(\alpha\gamma\delta^2 - \beta^2\delta^2 - \beta\gamma^2\delta + \gamma^4) \\ & + 3A^2 C(2\beta^2\gamma\delta - \alpha\gamma^2\delta - \beta\gamma^3) + 12AB^2(2\beta^2\gamma\delta - \alpha\gamma^2\delta - \beta\gamma^3) \\ & + 3C(AC + 4B^2)(\alpha\beta^2\delta + \beta^3\gamma - 2\alpha\beta\gamma^2) + 4B(2B^2 + 3AC)(\alpha\gamma^3 - \beta^3\delta) \\ & + 6BC^2(\alpha^2\gamma^2 + \alpha\beta^2\gamma - \alpha^2\beta\delta - \beta^4) + C^3(\alpha^3\delta + 2\alpha\beta^3 - 3\alpha^2\beta\gamma). \end{aligned}$$

M is called a skew invariant: because g is odd M will change sign if the transformation involves a reflection (i.e. the modulus of the transformation Δ is negative). M is related to the other invariants as follows [61]:

$$M^2 = -4Q^3 P^2 + P(R^2 + 12RQI + 24Q^2 I^2) - 4RI^3 - 36QI^4.$$

Since we have 7 coefficients we would expect to find $7 - 4 = 3$ independent absolute invariants, and hence 4 independent relative invariants — the simplest are Q, P, I and R.

The quartic form

Invariants:
$$\begin{aligned} S &= ae - 4bd + 3c^2, & g &= 4; \\ T &= ace + 2bcd - ad^2 - eb^2 - c^3 & g &= 6. \end{aligned}$$

The system of quartic and quadratic forms

Invariants additional to those of the quartic (S, T) and quadratic (Q) above:

$$L = eA^2 + 4cB^2 + aC^2 - 4bBC + 2cAC - 4dAB, \qquad g = 4$$

$$N = A^2(ce - d^2) + B^2(ae - c^2) + C^2(ac - b^2) \\ + 2BC(bc - ad) + 2AC(bd - c^2) + 2AB(cd - be), \quad g = 6.$$

A skew invariant also exists, whose square is a polynomial in the other invariants (i.e. it is irreducible but not independent):

$$H = A^3(3cde - 2d^3 - be^2) + A^2B(ae^2 + 2bde + 6cd^2 - 9c^2e) \\ + (AC + 4B^2)[A(3bce - ade - 2bd^2) - C(3acd - abe - 2b^2d)] \\ + 2B(2B^2 + 3AC)(ad^2 - b^2e) + BC^2(9ac^2 - a^2e - 2abd - 6b^2) \\ + C^3(a^2d + 2b^3 - 3abc), \qquad g = 9.$$

The system of quartic and cubic forms

The system has 17 irreducible invariants in addition to P, S and T; most of these will involve a large number of terms (see table 3.1). One of the simplest is

$$K = a(\beta\delta - \gamma^2)^2 - 2b(\alpha\delta - \beta\gamma)(\beta\delta - \gamma^2) - 2d(\alpha\gamma - \beta^2)(\alpha\delta - \beta\gamma) \\ + c\left[2(\alpha\gamma - \beta^2)(\beta\delta - \gamma^2) + (\alpha\delta - \beta\gamma)^2\right] + e(\alpha\gamma - \beta^2)^2, \qquad g = 8.$$

Another invariant is the skew invariant:

$$Z = \alpha^2(be^2 + 2d^3 - 3cde) + \alpha\beta(9c^2e - ae^2 - 2bde - 6cd^2) \\ + (2\alpha\gamma + 3\beta^2)(ade + 2bd^2 - 3bce) + (\alpha\delta + 9\beta\gamma)(b^2e - ad^2) \\ + (2\beta\delta + 3\gamma^2)(3acd - abe - 2b^2d) + \gamma\delta(a^2e + 2abd + 6b^2c - 9ac^2) \\ + \delta^2(3abc - a^2d - 2b^3), \qquad g = 9.$$

The system of quadratic, cubic and quartic forms

The whole system has a total of 65 irreducible invariants (see table 3.1), of which 38 contain coefficients of all three forms, and 38 are skew (but not the same 38!). The simplest invariant of this system to contain coefficients of all three forms is

$$G_1 = (Ac - 2Bb + Ca)(\beta\delta - \gamma^2) - (Ad - 2Bc + Cb)(\alpha\delta - \beta\gamma) \\ + (Ae - 2Bd + Cc)(\alpha\gamma - \beta^2), \qquad g = 6.$$

The next simplest is

$$G_2 = A(\alpha\delta d + 3\beta\gamma d + 4\gamma\delta b - \alpha\gamma e - \delta^2 a - 3\gamma^2 c - 3\beta\delta c) \\ + 2B(6\beta\gamma c + \alpha\beta e - 3\beta^2 d - \beta\delta b - 3\gamma^2 b - \alpha\gamma d) \\ + C(\alpha\delta b + 3\beta\gamma b + 4\alpha\beta d - \alpha^2 e - \beta\delta a - 3\alpha\gamma c - 3\beta^2 c), \qquad g = 6 (3.13)$$

which is equal to $a^{ij}b^{klm}b^{nop}c^{qrst}\epsilon_{il}\epsilon_{kj}\epsilon_{mr}\epsilon_{ot}\epsilon_{qn}\epsilon_{sp}$.

The quintic form

$ax^5 + 5bx^4y + \ldots + fy^5$ has four irreducible invariants, of order 4, 8, 12 and 18; the latter invariant is a skew invariant and is a function of the first three. The simplest of the invariants is

$$J = a^2f^2 - 10abef + 4acdf + 16ace^2 - 12ad^2e + 16b^2df + 9b^2e^2 \\ - 12bc^2f - 76bcde + 48bd^3 + 48c^3e - 32c^2d^2, \qquad g = 10.$$

The sextic form

$ax^6 + 6bx^5 + \ldots gy^6$ has five irreducible invariants, four of which are independent and one of which is skew. The simplest two are

$$
\begin{aligned}
E &= ag - 6bf + 15ce - 10d^2, \qquad g = 6. \\
F &= aceg - acf^2 - ad^2g + 2adef - ae^3 - b^2eg + b^2f^2 + 2bcdg - 2bcef - 2bd^2f \\
&\quad + 2bde^2 - c^3g + 2c^2df + c^2e^2 - 3cd^2e + d^4, \qquad g = 12.
\end{aligned}
$$

3.4.2 Invariants of ternary forms

Invariants of five lines

As we saw in the first section of this chapter, equation (3.12), five coplanar lines defined as $a_ix + b_iy + c_i = 0$, have an invariant given by [9]:

$$
(I_{431}I_{521})/(I_{421}I_{531}),
$$

where I_{ijk} is the determinant of the matrix $[\, l_i \quad l_j \quad l_k \,]$, with $l_i^T = [\, a_i \quad b_i \quad c_i \,]$.

Invariants of two conics

A conic is a quadratic plane curve defined by $Ax^2 + Bxy + Cy^2 + Dx + Ey + F = 0$, and a pair of conics have two invariants I_{C1} and I_{C2} based on the matrices C_i, $i = 1, 2$ [73]:

$$
C_i = \begin{bmatrix} A_i & B_i/2 & D_i/2 \\ B_i/2 & C_i & E_i/2 \\ D_i/2 & E_i/2 & F_i \end{bmatrix};
$$

$$
I_{C1} = \frac{\text{Tr}(C_1^{-1}C_2)\,|C_1|}{\left[\text{Tr}(C_2^{-1}C_1)\right]^2 |C_2|}; \qquad I_{C2} = \frac{\text{Tr}(C_2^{-1}C_1)\,|C_2|}{\left[\text{Tr}(C_1^{-1}C_2)\right]^2 |C_1|}; \tag{3.14}
$$

where $\text{Tr}(C_i)$ is the trace of C_i.

Invariant of a conic and two lines

Similarly, a conic C and two lines $l_i^T = [\, a_i \quad b_i \quad c_i \,]$, $i = 1, 2$, has a single joint invariant I_{Cl} [74]:

$$
I_{Cl} = \frac{\left(l_1^T C^{-1} l_2\right)^2}{(l_1^T C^{-1} l_1)\,(l_2^T C^{-1} l_2)}.
$$

Invariants of a ternary quartic

The simplest two invariants of a ternary quartic are listed by Salmon, pp. 263-267 [60], and are repeated below for convenience. If we write the ternary quartic as

$$
\begin{aligned}
&ax^4 + by^4 + cz^4 + 6fy^2z^2 + 6gx^2z^2 + 6hx^2y^2 + 12lx^2yz + 12mxy^2z + 12nxyz^2 \\
&+ 4a_2x^3y + 4a_3x^3z + 4b_1xy^3 + 4b_3y^3z + 4c_1xz^3 + 4c_2yz^3 = 0,
\end{aligned}
$$

then the simplest invariants A and B are given by

$$
\begin{aligned}
A = {} & abc + 3(af^2 + bg^2 + ch^2) - 4(ab_3c_2 + bc_1a_3 + ca_2b_1) + 12(fl^2 + gm^2 + hn^2) \\
& + 6fgh - 12lmn - 12(a_2nf + a_3mf + b_1ng + b_3lg + c_1mh + c_2lh) \\
& + 12(lb_1c_1 + mc_2a_2 + na_3b_3) + 4(a_2b_3c_1 + a_3b_1c_2);
\end{aligned}
\tag{3.15}
$$

$$
B = \begin{vmatrix}
a & h & g & l & a_3 & a_2 \\
h & b & f & b_3 & m & b_1 \\
g & f & c & c_2 & c_1 & n \\
l & b_3 & c_2 & f & n & m \\
a_3 & m & c_1 & n & g & l \\
a_2 & b_1 & n & m & l & h
\end{vmatrix}
$$

(B is the determinant of the matrix). Since A has weight $g = 4$ and B has weight $g = 8$, A^2/B is an absolute invariant.

3.4.3   Other functions invariant to planar projection

The theory of algebraic invariants of ternary forms considers functions of the form's coefficients; one can also form invariant image features using covariants, if one uses the coordinates of reference points in the image.

Invariant of five points

Given five points with coordinates (x_i, y_i), one can form an invariant based on the invariant of five lines by redefining I_{ijk} to contain $\mathbf{v}_i^T = [x_i \; y_i \; 1]$ instead of $\mathbf{l}_i^T$. This invariant can also be interpreted as a cross-ratio of areas of triangles, as we will see in chapter 5.

Invariant of a conic and two points

A conic $\mathbf{C}$ and two points $\mathbf{v}_i^T = [\; x_i \; y_i \; z_i \;]$ in homogeneous coordinates, $i = 1, 2$, has a single joint invariant I_{Cp}:

$$
I_{Cp} = \frac{\left(\mathbf{v}_1^T \mathbf{C} \mathbf{v}_2\right)^2}{\left(\mathbf{v}_1^T \mathbf{C} \mathbf{v}_1\right)\left(\mathbf{v}_2^T \mathbf{C} \mathbf{v}_2\right)}.
\tag{3.16}
$$

Invariant of two points and two lines

Weiss *et al.* [74] list the following invariant:

$$
I_{lp} = \frac{\mathbf{l}_1^T \mathbf{v}_1}{\mathbf{l}_2^T \mathbf{v}_1} \cdot \frac{\mathbf{l}_2^T \mathbf{v}_2}{\mathbf{l}_1^T \mathbf{v}_2}.
$$

They argue that the invariant should not exist, because two points and two lines on a plane have eight parameters, and one needs at least nine to obtain an absolute invariant of an eight parameter transformation. However, one can see that it is easy to construct four collinear points by drawing a line through the two points and seeing where it intersects the two lines. As a result we see that the invariant only contains information along the line through the two points; the 'missing parameter' contains scale information perpendicular to the line.

3.5 Conclusions

We have looked at invariants of binary forms and of ternary forms; the former play a crucial role in allowing one to use moments to form features invariant to affine transformations, while the latter are important in obtaining invariance to projective transformations.

We saw that the quadratic, cubic and quartic binary forms have nine independent invariants, but also a large number of irreducible invariants. Many of the irreducible invariants are high order polynomials in the coefficients with a large number of terms, which means that the moment invariants based on them are more susceptible to image distortions, as will be demonstrated in the following chapter. Nevertheless, we will see that a number of moment invariants are robust and useful features, in particular when recognizing partially occluded objects in chapter 6.

A ternary form defines a polynomial plane curve, from which we saw that polynomial plane curves of a given order remain polynomial curves of the same order under planar projection. To obtain invariant functions of the polynomial's coefficients one needs more coefficients than degrees of freedom (eight); hence five lines provide two independent invariants, as do two conics; a cubic provides one and a quartic provides six, one of which was listed explicitly. Chapter 5 discusses how these invariants can be used to recognize projectively transformed images.

Chapter 4

Invariance to Affine Transformations

4.1　Introduction

This chapter is devoted to obtaining features invariant to affine image transformations. It starts with the fundamental theorem of moment invariants, introduced by Hu in 1962 [26] and recently corrected by the author [27], which shows how one can obtain invariant functions of the image moments directly from 2D algebraic invariants. In addition to stating the revised fundamental theorem of moment invariants (RFTMI), section 4.2 lists a number of affine moment invariants and contrast moment invariants, discusses their robustness to noise and gives a novel presentation of the effects of symmetries. The next section discusses an alternative approach based on moments, normalization, which was first discussed for affine transformations by Udagawa in 1964 [47] and solved fully by Dirilten & Newman in 1977 [75]. As we shall see, the main drawback with normalization is its vulnerability to rotational symmetries in the object, making moment invariants more useful. Section 4.4 covers a number of alternative techniques that do not use the image's moments: Fourier descriptors, introduced by Arbter *et al.* [3]; differential methods based on signatures [6, 22] and point-based methods [19], and a novel method based on correlations devised by the author.

The results of six experiments are presented in section 4.3. The first part discusses experiments that have been described in the literature on the subject, before describing experiments performed by the author comparing moment invariants with Fourier descriptors for coarsely sampled images. The second experiment compares moment invariants with correlation invariants for a larger number of coarsely sampled images; in both cases the moment invariants perform significantly better. The next experiment investigates the performance of moment invariants with grey-level images under different lighting conditions; normalization against brightness and contrast is seen to be necessary, resulting in good performance by some of the moment invariants. The fourth experiment investigates the ability of moment invariants to detect reflectional and rotational symmetries, and finds that the latter can easily be detected but not the former. The last two experiments investigate the stability of the point-based invariants when one of the points is perturbed, and the stability of the moment invariants when viewing coarsely sampled near-planar objects with noticeable perspective distortion. In the latter case, the moment invariants are fairly stable and provide good discrimination.

Under weak perspective, the parameters of an image's affine transformation can be used to express the motion of the planar object in 3D, a fact which can be useful when verifying hypotheses as discussed in chapter 6. The equations relating 3D motion to an affine transformation are given in appendix C.

4.2   Moment invariants

4.2.1   The revised fundamental theorem of moment invariants

Invariants to affine image transformations can easily be constructed from algebraic invariants by using the revised fundamental theorem of moment invariants (RFTMI) [27]; the original theorem was introduced by Hu in 1962 [26] but contains an error which results in his invariants to affine image transformations being incorrect. Below the theorem is stated; a proof can be found in appendix B.

The Revised Fundamental Theorem of Moment Invariants
Let $|\Delta|$ be the absolute value of the determinant Δ of the image transformation. Then, if the binary form of order p has an algebraic invariant $I(a_{p,0}, a_{p-1,1}, \ldots, a_{0,p})$ of weight g and order k, i.e.

$$I(a'_{p,0}, a'_{p-1,1}, \ldots, a'_{0,p}) = \Delta^g I(a_{p,0}, a_{p-1,1}, \ldots, a_{0,p}),$$

then the central moments of order p have the same invariant but with the additional factor $|\Delta|^k$:

$$I(\mu'_{p,0}, \mu'_{p-1,1}, \ldots, \mu'_{0,p}) = \Delta^g |\Delta|^k I(\mu_{p,0}, \mu_{p-1,1}, \ldots, \mu_{0,p}). \tag{4.1}$$

(The theorem as stated by Hu [26] has $k \equiv 1$.) The theorem also holds for algebraic invariants containing coefficients from two or more binary forms of different orders.

The following example shows how the RFTMI is applied: from the previous section we know that $I(a_{20}, a_{11}, a_{02}) = a_{20}a_{02} - a_{11}^2$ is an algebraic invariant of $a_{20}x^2 + 2a_{11}xy + a_{02}y^2$ with weight $w = 2$. I is a second order polynomial in a_{20}, a_{11} and a_{02} so $k = 2$. The RFTMI tells us that

$$I(\mu'_{20}, \mu'_{11}, \mu'_{02}) = \Delta^2 |\Delta|^2 I(\mu_{20}, \mu_{11}, \mu_{02}), \tag{4.2}$$

i.e. $$\mu'_{20}\mu'_{02} - \mu'^2_{11} = \Delta^2 |\Delta|^2 (\mu_{20}\mu_{02} - \mu_{11}^2). \tag{4.3}$$

As in the case of algebraic invariants, absolute invariants can be obtained by using the ratio of relative invariants. If we set $\mu = \mu_{00}$ and $\mu' = \mu'_{00}$, then

$$\mu' = |\Delta|\mu. \tag{4.4}$$

If the weight g of the invariant I is divisible by 2, $\Delta^g = |\Delta|^g$ and we can use (4.4) to obtain the absolute invariant

$$\frac{I'}{\mu'^{g+k}} = \frac{|\Delta|^{g+k}I}{|\Delta|^{g+k}\mu^{g+k}} = \frac{I}{\mu^{g+k}}. \tag{4.5}$$

As an example let $I = I(\mu_{20}, \mu_{11}, \mu_{02}) = \mu_{20}\mu_{02} - \mu_{11}^2$ as given in (4.2) and (4.3), so $I' = |\Delta|^4 I$ and

$$\frac{I'}{\mu'^4} = \frac{|\Delta|^4 I}{(|\Delta|\mu)^4} = \frac{I}{\mu^4}.$$

If g is not divisible by 2 then the invariant I/μ^{g+k} is a skew invariant — its sign depends on whether $\Delta > 0$ or $\Delta < 0$:

$$\frac{I'}{\mu'^{g+k}} = \frac{\Delta|\Delta|^{g-1+k}I}{|\Delta|^{g+k}\mu^{g+k}} = \frac{\Delta I}{|\Delta|\mu^{g+k}} = \text{sign}(\Delta)\frac{I}{\mu^{g+k}}.$$

where $\text{sign}(x) = +1$ if $x > 0$ and $\text{sign}(x) = -1$ if $x < 0$. Skew invariants using central moments are zero for objects with symmetry under reflection; this is discussed further in section 4.5.

Bamieh & de Figueiredo [76] and the author [67] independently show how to obtain invariants by contracting the moment tensor introduced by Cyganski & Orr [58] to perform normalization; Bamieh & de Figueiredo's treatment contains a slight error however, which denies the existence of the skew invariants. Generating moment invariants directly using tensors is discussed in the last section of appendix B. Recently, Taubin & Cooper [77], apperently unaware of the above papers, have rediscovered moment invariants for affine transformations using matrix techniques; these are, however, clumsier than the use of tensors.

4.2.2 Examples of image invariants

Invariance to affine image transformations

We obtain further features invariant to affine image transformations by proceeding as above, where we replace the coefficients $(a, b, c, \ldots)$ in the expressions for the algebraic invariants by the corresponding central moments i.e. the coefficients A, B and C of the binary form $(A, B, C)(x, y)^2$ are replaced by μ_{20}, μ_{11} and μ_{02} respectively; similarly, the coefficients α, β, γ and δ of the cubic form $(\alpha, \beta, \gamma, \delta)(x, y)^3$ are replaced by μ_{30}, μ_{21}, μ_{12} and μ_{03} respectively, and so on for higher order forms. As an example, the simplest invariant Q becomes $\mu_{20}\mu_{02} - \mu_{11}^2$. If we use only central moments of up to fourth order, we have 13 non-zero moments, so we expect to find $13 - 4 = 9$ independent absolute invariants. One set of nine is presented below as $\psi_1, \ldots, \psi_9$, where $\mu = \mu_{00}$ (ψ_1, ψ_2, ψ_3 and ψ_4 first appeared in [27]):

$$\psi_1 = \frac{Q}{\mu^4}; \qquad \psi_2 = \frac{P}{\mu^{10}}; \qquad \psi_3 = \frac{I}{\mu^7};$$

$$\psi_4 = \frac{R}{\mu^{11}}; \qquad \psi_5 = \frac{S}{\mu^6}; \qquad \psi_6 = \frac{T}{\mu^9}; \qquad (4.6)$$

$$\psi_7 = \frac{L}{\mu^7}; \qquad \psi_8 = \frac{N}{\mu^{10}}; \qquad \psi_9 = \frac{G_1}{\mu^{10}}.$$

We also have a large number of irreducible invariants (see Table 3.1) which, although not independent, could still prove useful for classification - remember that the sign of irreducible invariants cannot be determined from the independent ones. As

Invariant:	μ	Q	P	I	R	S	T	L	N	K	G_i	J	E	F
$g+k$	1	4	10	7	11	6	9	7	10	13	10	14	8	16
k	1	2	4	3	5	2	3	3	4	5	4	4	2	4

Table 4.1: The values of $g+k$ and k for the non-skew invariants presented earlier.

can be seen from the table, most of the irreducible invariants are of quite a high order in the coefficients, and hence will have a large number of terms in their expressions. This is undesirable primarily because they will be more noise sensitive than invariants with fewer terms — see section 4.2.3.

An example of an irreducible absolute invariant is

$$\psi_{10} = \frac{K}{\mu^{13}}.$$

Three of the absolute (irreducible) skew invariants are

$$\chi_1 = \frac{M}{\mu^{16}}; \qquad \chi_2 = \frac{Z}{\mu^{14}}; \qquad \chi_3 = \frac{H}{\mu^{15}}.$$

We can also form absolute invariants using fifth and sixth order moments; the simplest are

$$\Psi_1 = \frac{J}{\mu^{14}}; \qquad \Psi_2 = \frac{E}{\mu^8}; \qquad \Psi_3 = \frac{F}{\mu^{16}}.$$

Invariance to changes in contrast and affine image transformations

A change in contrast by a factor c plus an affine coordinate transformation results in a new image $\hat{f}(\hat{x}, \hat{y})$ related to the original by $f(x, y) = c\hat{f}(\hat{x}, \hat{y})$ [40, 27]. It is easy to see from the definition of moments that the central moments $\hat{\mu}_{pq}$ of $\hat{f}(\hat{x}, \hat{y})$ are related to those of $f(x, y)$ by $\mu_{pq} = c|\Delta|\hat{\mu}_{pq}$. This in turn means that an invariant function $I(.)$ of the $\hat{\mu}_{pq}$ is related to the same function of the μ_{pq} by

$$I(\mu_{p0}, \ldots, \mu_{0p}) = \Delta^g |\Delta|^k c^k I(\hat{\mu}_{p0}, \ldots, \hat{\mu}_{0p}). \qquad (4.7)$$

If we ignore the skew invariants, so g is even, we have $I(\mu'_{p0}, \ldots) = |\Delta|^{g+k} c^k I(\mu_{p0}, \ldots)$. Absolute invariants can be constructed by dividing combinations of invariants by other combinations and thus eliminating the factors $|\Delta|$ and c. Table 4.1 lists the values of $g+k$ and k for the non-skew invariants presented earlier. Many different combinations are possible; below are a few examples (the first three are also presented in [27]):

$$\Gamma_1 = \frac{R}{\mu P}; \qquad \Gamma_2 = \frac{Q^2}{\mu I}; \qquad \Gamma_3 = \frac{QI}{R};$$

$$\Gamma_4 = \frac{\mu S}{I}; \qquad \Gamma_5 = \frac{\mu T}{P}; \qquad \Gamma_6 = \frac{L}{I};$$

$$\Gamma_7 = \frac{N}{P}; \qquad \Gamma_8 = \frac{K}{IS}; \qquad \Gamma_9 = \frac{G_i}{P};$$

$$\Gamma_{10} = \frac{JQ}{T^2}; \qquad \Gamma_{11} = \frac{\mu^2 E}{P}; \qquad \Gamma_{12} = \frac{F}{E^2}.$$

To see that Γ_1 is an invariant, use Table 4.1 and (4.7):

$$\Gamma_1 = \frac{R}{\mu P} = \frac{|\Delta|^{11} c^5 \hat{R}}{|\Delta| c \hat{\mu} |\Delta|^{10} c^4 \hat{P}} = \frac{\hat{R}}{\hat{\mu} \hat{P}} = \hat{\Gamma}_1.$$

We saw in chapter 2 that one can easily achieve invariance to changes in brightness and contrast by suitably normalizing a grey-level image, so one might ask oneself what purpose the above contrast invariants serve. In fact, they can be useful additional features even when classifying binary images where no changes in contrast occur: since they are nonlinear functions of other invariants, they can provide discrimination when others fail, a fact which is borne out by the experiment described in section 4.5.1.

4.2.3 Robustness to noise

As we saw in chapter 2, Abu-Mostafa & Psaltis [36] have shown that the signal to noise ratio of the moment m_{pq} of a continuous image subject to additive white noise is proportional to $1/\sqrt{p+q}$ i.e. moments become more susceptible to noise as their order increases. It is also easy to show that moment invariants are biased estimators in the presence of noise: if we assume each measured moment m_{pq} is made up of an underlying, correct value $\hat{m}_{pq}$ plus a zero-mean perturbation n_{pq}: $m_{pq} = \hat{m}_{pq} + n_{pq}$, then the measured moments are unbiased: $E[m_{pq}] = \hat{m}_{pq}$, but any polynomial in them will bring in cross-product terms, resulting in bias. For example,

$$E[Q] = E\left[\mu_{20}\mu_{02} - \mu_{11}^2\right] = \hat{\mu}_{20}\hat{\mu}_{02} - \hat{\mu}_{11}^2 - \sigma_{11}^2 = \hat{Q} - \sigma_{11}^2,$$

where σ_{11}^2 is the variance of μ_{11}. Clearly, the higher the order of the polynomial in the moments and the larger the number of terms in the invariant, the worse the bias will be. Similarly, the variance of a moment invariant increases with order and number of terms, so one is advised to use the simplest invariants. However, since one is likely to obtain better discrimination by using higher order invariants, we see that there is a limit to the discriminatory power of moment invariants, and that it is governed by the noise (both additive and due to quantization effects) on the image; in most cases the quantization noise is more significant than additive noise. It is this limit that led Abu-Mostafa & Psaltis [36] to recommend the use of image normalization followed by template matching, as discussed in section 4.3. We will see, however, that normalization also has its disadvantages — in particular, one loses the ability to index into a database.

4.2.4 The effects of object symmetries

We have already come across skew invariants, which are zero if an object has reflectional symmetry, or can be affinely transformed to obtain an image with reflectional symmetry. This property follows from the fact that, if I_s is a skew invariant, then a reflection will result in $-I_s$. If an object has reflectional symmetry, a reflection leaves it unchanged, so $-I_s = I_s$. i.e. $I_s = 0$. The simplest skew invariants listed in section 3.4.1 are quite high order polynomials, so the previous section's discussion of robustness implies that they may not be very robust. Unfortunately, the experiments discussed in section 4.5 show that this is indeed the case when using coarsely sampled images: the skew invariants are unable to discriminate between two objects where one is the reflection of the other.

Skew invariants are zero if the object has a reflectional symmetry under affine transformations; similarly, any invariant which is a function of odd-order moments will be zero if an image has even order rotational symmetry under affine transformation. To see this, consider the moments m_{pq} of an image $f(x, y)$ with two-fold rotational symmetry i.e. $f(x, y) = f(-x, -y)$:

$$m_{pq} = \int_{-\infty}^{+\infty} \int_{-\infty}^{+\infty} x^p y^q f(x, y) \, dx \, dy.$$

A rotation of the axes by 180° corresponds to $x = -x'$, $y = -y'$, giving

$$m_{pq} = \int_{-\infty}^{+\infty} \int_{-\infty}^{+\infty} (-x')^p (-y')^q f(-x', -y') \, dx' \, dy'.$$

If $p + q$ is odd, and $f'(x', y') = f(-x', -y')$ is the rotated image,

$$m_{pq} = - \int_{-\infty}^{+\infty} \int_{-\infty}^{+\infty} x'^p y'^q f'(x', y') \, dx' \, dy' = -m'_{pq}.$$

However, if $f(x, y)$ has twofold rotational symmetry, then $m_{pq} = m'_{pq}$, which means that $m_{pq} = -m_{pq}$, and hence $m_{pq} = 0$.

Since moment invariants are homogeneous polynomials, a moment invariant which is a function of third order moments will contain a third order moment in each term (c.f. P, I, R and M in section 3.4.1), and hence will be zero if an object has two-fold rotational symmetry. Since a moment invariant is invariant to affine transformations, it will be zero for any affine transformation of an object with two-fold rotational symmetry.

The experiments discussed in section 4.5 show that the invariants distinguish between the presence or absence of rotational symmetry much more accurately than they distinguish between that of reflectional symmetry.

4.3   Normalization using moments

This section presents an alternative use of moments to obtain features invariant to affine transformations, namely normalization. As we shall see, normalization is made difficult if an object has any rotational symmetries.

An affine transformation has six parameters — two for the translation and four for the general linear transformation. Hence an affine transformation and a change in contrast has seven parameters. Dirilten & Newman [75] show that transforming the image to satisfy the following six relationships $\mu_{00} = \mu_{20} = \mu_{02} = 1$; $\mu_{10} = \mu_{01} = \mu_{11} = 0$ reduces the seven degrees of freedom to one, that one being that of the orthogonal transformations (rotation and/or reflection). Doing this allows one to deal with rotational symmetry in a constructive manner, something which is not possible if one solves Cyganski & Orr's [58, 78] linear equations for the normalizing parameters. Reeves *et al.* [4] use essentially the same criteria, but $\mu_{20} = \mu_{02}$ is not given a predetermined value because they only deal with binary images. Leu [79] has rediscovered the above, but in a rather ad hoc manner and without mentioning the problems of rotational symmetry. Note that the normalization equations can also be used to find the affine transformation linking two views of an object, an approach developed by Cyganski & Orr [58]. Using the equations of appendix C then allows one to compute the object's motion in 3D.

Dirilten & Newman use group theory to prove the above, but one can also do so using the definition of moments; in addition to giving a new perspective on how the normalization works, it allows us to explicitly write down the normalizing equation in terms of the unnormalized image moments. Having demonstrated Dirilten & Newman's result, we will go on to see how to normalize against the rotations and reflections of the semi-normalized image.

The following notation will be used below:

m_{pq} – The regular moments of the input image.

μ_{pq} – The central moments of the input image.

η_{pq} – The central moments of the semi-normalized image, obtained after normalizing using step 1.

ν_{pq} – The central moments of the image normalized against all affine transformations except reflection.

c_{pq} – The complex moments of the normalized image with central moments ν_{pq}.

Step 1   Find the contrast factor C and the linear transformation $x' = ax + by$, $y' = cx + dy$, that maps the input image into a semi-normalized version, invariant to all bar orthogonal transformations.

As defined above, $\{\mu_{pq}\}$ are the central moments of the input image $f(x, y)$ and $\{\eta_{pq}\}$ those of the semi-normalized image $f'(x', y')$ where $f'(x', y') = Cf(x, y)$. The Jacobian J of the above linear transformation is $J = ad - bc$. Consider η_{pq}:

$$\eta_{pq} = \int_{-\infty}^{\infty}\int_{-\infty}^{\infty} x'^p y'^q f'(x', y')\, dx'\, dy'$$

$$= C\int_{-\infty}^{\infty}\int_{-\infty}^{\infty} (ax + by)^p (cx + dy)^q f(x, y)|J|\, dx\, dy. \tag{4.8}$$

This gives us

$$\eta_{00} = C|J|\mu_{00}; \tag{4.9}$$

$$\eta_{20} = C|J|(a^2\mu_{20} + 2ab\mu_{11} + b^2\mu_{02}); \tag{4.10}$$

$$\eta_{11} = C|J|(ac\mu_{20} + (ad + bc)\mu_{11} + bd\mu_{02}); \tag{4.11}$$

$$\eta_{02} = C|J|(c^2\mu_{20} + 2cd\mu_{11} + d^2\mu_{02}). \tag{4.12}$$

Since we are using central moments, the six constraints of Dirilten & Newman become four: $\eta_{00} = \eta_{20} = \eta_{02} = 1$ and $\eta_{11} = 0$. We have five parameters a, b, c, d and C and only four constraints, so we will be able to choose the value of one of the parameters arbitrarily. The constraints $\eta_{00} = 1$ and $\eta_{20} = 1$ together with (4.9) and (4.10) give us:

$$a^2\mu_{20} + 2ab\mu_{11} + b^2\mu_{02} = \mu_{00}.$$

If $\mu_{20} \neq 0$ we can choose $b = 0$, giving us:

$$b = 0 \;\Rightarrow\; a = \left(\frac{\mu_{00}}{\mu_{20}}\right)^{\frac{1}{2}}. \tag{4.13}$$

If $\mu_{20} = 0$, we use the constraint $\eta_{02} = 1$ and (4.12), and choose $c = 0$:

$$c = 0 \;\Rightarrow\; d = \left(\frac{\mu_{00}}{\mu_{02}}\right)^{\frac{1}{2}}.$$

If we assume $\mu_{20} \neq 0$, the constraints $\eta_{11} = 0$ and $\eta_{02} = 1$ give us, using (4.11), (4.12) and (4.13):

$$a = \left(\frac{\mu_{00}}{\mu_{20}}\right)^{\frac{1}{2}}; \quad b = 0; \quad c = -\frac{\mu_{11}}{\mu_{20}}d; \quad d = \left(\frac{\mu_{00}\mu_{20}}{\mu_{20}\mu_{02} - \mu_{11}^2}\right)^{\frac{1}{2}};$$

$$C = \frac{1}{\mu_{00}^2}\left(\mu_{20}\mu_{02} - \mu_{11}^2\right)^{\frac{1}{2}}. \tag{4.14}$$

The normalizing transformation given above contains no rotation, because $b = 0$, and allows reflections because $ad > 0$ and $ad < 0$ are both equally possible.

Step 2   Find the rotation that will fully normalize the semi-normalized image produced by step 1 (except for a possible reflection).

We saw earlier that we must rotate the image until the complex moment c_{pq} is positive real. If we assume the image has no rotational symmetry, we can choose $p = 2$ and $q = 1$. Making c_{21} positive real is the same as making $\nu_{30} + \nu_{12} > 0$ and $\nu_{03} + \nu_{21} = 0$, where ν_{pq} are the moments of the fully normalized image. A rotation corresponds to a linear transformation $x' = \alpha x + \beta y$, $y' = \gamma x + \delta y$, where $\alpha = \delta$, $\beta = -\gamma$ and $\alpha^2 + \beta^2 = 1$. If we put these values into (4.8) we find that

$$\alpha = \frac{\eta_{12} + \eta_{30}}{\{(\eta_{12} + \eta_{30})^2 + (\eta_{21} + \eta_{03})^2\}^{\frac{1}{2}}}, \quad \beta = \frac{\eta_{21} + \eta_{03}}{\{(\eta_{12} + \eta_{30})^2 + (\eta_{21} + \eta_{03})^2\}^{\frac{1}{2}}}. \tag{4.15}$$

Hence the transformation to obtain an image normalized against affine transformations and changes in contrast is $f'(x', y') = C f(x, y)$,

$$\begin{bmatrix} x' \\ y' \end{bmatrix} = \begin{bmatrix} \alpha & \beta \\ -\beta & \alpha \end{bmatrix} \begin{bmatrix} a & 0 \\ c & d \end{bmatrix} \begin{bmatrix} x - \bar{x} \\ y - \bar{y} \end{bmatrix} \tag{4.16}$$

with a, c, d and C given in (4.14) and α, β given in (4.15); $\bar{x} = m_{10}/m_{00}$, $\bar{y} = m_{01}/m_{00}$, where the $\{m_{pq}\}$ are the regular moments of the input image. The normalized image satisfies $c_{00} = 1$, $c_{11} = 2$, $c_{00} = c_{20} = 0$, c_{21} positive real, where $\{c_{pq}\}$ are its complex moments.

Note that if we have a reference point (r_x, r_y) on the image, one can replace $\bar{x}$ with r_x and $\bar{y}$ with r_y in the above and use m_{10} and m_{01} to determine α and β:

$$\alpha = \frac{a\eta_{10}}{\{a^2\eta_{10}^2 + (c\eta_{10} + d\eta_{01})^2\}^{\frac{1}{2}}}; \quad \beta = \frac{c\eta_{10} + d\eta_{01}}{\{a^2\eta_{10}^2 + (c\eta_{10} + d\eta_{01})^2\}^{\frac{1}{2}}};$$

If we want to normalize against reflection too, we can stipulate that the imaginary part of c_{30} be positive after rotation normalization i.e. $3\nu_{21} - \nu_{03} > 0$, where the $\{\nu_{pq}\}$ are the moments of the image normalized using (4.16). To work out the condition in terms of the central moments μ_{pq} of the input image, we must use the linear transformation defined by A, B, C and D, where

$$\begin{bmatrix} A & B \\ C & D \end{bmatrix} = \begin{bmatrix} \alpha & \beta \\ -\beta & \alpha \end{bmatrix} \begin{bmatrix} a & 0 \\ c & d \end{bmatrix}.$$

i.e. $A = \alpha a + \beta c$; $B = \beta d$; $C = \alpha c - \beta a$; $D = \alpha d$.

Using (4.8) gives us

$$3\nu_{21} - \nu_{03} = B(B^2 - 3A^2)\mu_{30} + 3A(A^2 - 3B^2)\mu_{21} + 3B(3A^2 - B^2)\mu_{12} + A(3B^2 - A^2)\mu_{03}.$$

If this expression is positive, we use the normalization of (4.16); otherwise we add a reflection about the x-axis to the transformation in (4.16), which is the same as using the rotation plus reflection matrix

$$\begin{bmatrix} \alpha & \beta \\ \beta & -\alpha \end{bmatrix} \quad \text{in place of} \quad \begin{bmatrix} \alpha & \beta \\ -\beta & \alpha \end{bmatrix}.$$

We have so far assumed that none of the objects to be classified possess rotational symmetry. If this is not the case, one must test for the presence of such symmetries before deciding which complex moment c_{pq} should be made positive real in the normalization. For example, in the experiments on the ten digits in section 4.5, many of the digits possess two-fold rotational symmetry. If c_{21} is approximately zero, c_{31} should be made positive real; otherwise, use c_{21}. Clearly, the higher the degree of rotational symmetry, the higher the order of the moments required to normalize the image. In particular, one cannot use normalization to classify objects with more than four-fold rotational symmetry, such as hexagons, using moments of order up to four. Furthermore, circles and ellipses cannot be normalized at all — one would instead have to test for circular symmetry one order higher than that possessed by the second-most symmetric object being classified.

4.3.1 Moments of the normalized image

Udagawa [47] first pointed out that one can obtain features invariant to affine transformations by using the moments of the normalized image, and Reeves *et al.* [4] have performed some experiments using them. They can easily be computed from the central moments η_{pq} of the input image by inserting the normalizing parameters a, b, c, d and C into (4.8) above. Since the moments up to second order are already fixed by the normalization, the simplest invariant features are the third order moments. For a centred image,

$$\mu_{pq} = \int_{-\infty}^{+\infty} \int_{-\infty}^{+\infty} x^p y^q f(x,y) \, dx \, dy.$$

Using (4.8) we can write ν_{30} as

$$\begin{aligned} \nu_{30} &= C \int_{-\infty}^{+\infty} \int_{-\infty}^{+\infty} (ax + by)^3 f(x,y) \, dx \, dy. \\ &= C|\Delta|(a^3\mu_{30} + 3a^2 b\mu_{21} + 3ab^2 \mu_{12} + b^3 \mu_{03}). \end{aligned}$$

Similarly for the other moments:

$$\nu_{21} = C|\Delta|(a^2c\mu_{30} + (a^2d + 2abc)\mu_{21} + (2abd + b^2c)\mu_{12} + b^2d\mu_{03});$$
$$\nu_{12} = C|\Delta|(ac^2\mu_{30} + (bc^2 + 2acd)\mu_{21} + (ad^2 + 2bcd)\mu_{12} + bd^2\mu_{03});$$
$$\nu_{03} = -\nu_{21} \text{ by the normalization constraint;}$$

4.4 Alternative methods

Above we have seen two different approaches to obtaining affine invariants using image moments; both are so-called *global* methods, because they use information contained in all parts of the image. This section discusses alternative techniques, starting with a global one (Fourier descriptors), then summarizing *local* and *semi-local* invariants in the section on differential methods, so-called because they use derivatives of the object's boundary, before concluding with a novel global method based on correlations.

4.4.1 Fourier descriptors

Affine invariant Fourier descriptors were introduced by Arbter *et al.* [3, 80] for binary images, and have recently been generalized to cope with grey level images [17, 81]. Fourier descriptors were originally introduced to provide rotation invariance: if one has a closed curve (object boundary) described by $(x(s), y(s))$, $s \in [0, S]$, then the curve can be approximated by a Fourier series with coefficients U_k, V_k defined as

$$[U_k \ V_k] = \frac{1}{S} \int_0^S [x(s) \ y(s)] \ e^{-j\frac{2\pi ks}{S}} \ ds.$$

The magnitudes of U_k and V_k are invariant to rotations; invariance to translations can be achieved by placing the coordinate origin at the image centroid, and invariance to changes in scale by forming the ratio of two coefficients. Invariance to affine transformations is not so straightforward because the curve length can change; Arbter *et al.* [3] get around this by reparameterizing the arc-length of the boundary curve, having first placed the coordinate origin at the centroid: the arc-length

$$t = \frac{1}{2} \int_0^S |x(s)\dot{y}(s) - y(s)\dot{x}(s)| \ ds \tag{4.17}$$

is a relative invariant under linear transformations, where the original curve has coordinates $(x(s), y(s))$, and $\dot{x}(s)$ is the derivative of $x(s)$ with respect to s. Arbter *et al.* show that t is simply the swept area obtained by starting at the starting point $(x(0), y(0))$ and following the curve to the point $(x(S), y(S))$.

If the linear part of the affine transformation is represented by the 2×2 matrix $\mathbf{A}$, then

$$\begin{bmatrix} \hat{x} \\ \hat{y} \end{bmatrix} = \mathbf{A} \begin{bmatrix} x \\ y \end{bmatrix} \qquad \Rightarrow \qquad \begin{bmatrix} \hat{U}_k \\ \hat{V}_k \end{bmatrix} = \mathbf{A} \begin{bmatrix} U_k \\ V_k \end{bmatrix}.$$

If one chooses a different starting point on the boundary of the object viewed in a different position, the parameterization changes as follows: $\hat{t} = ct + d$, which in turn affects the Fourier coefficients as follows: $[\hat{U}_k \ \hat{V}_k]^T = \exp(jb_k)[U_k \ V_k]$, where b_k is a constant proportional to k. To obtain an affine invariant, one must eliminate $\mathbf{A}$

and $\exp(jb_k)$. Arbter *et al.* [3] pose the problem as that of finding solutions to a diophantine equation, of which the first order solution is given by

$$Q_k = \frac{U_k V_p^* - V_k U_p^*}{U_p V_p^* - V_p U_p^*}, \qquad k \neq p, \quad p \text{ fixed.}$$

(V_p^* is the complex-conjugate of V_p.) $|Q_k|$ is invariant to affine transformations (p is usually set to 1, see reference [3]); in fact, it is also invariant to changes in contrast. Recently, Jin & Yan [82] have proposed an alternative, which they claim is more robust: $I_k = E_k/E_1$, where

$$E_k = \begin{vmatrix} U_k^r & U_k^i \\ V_k^r & V_k^i \end{vmatrix}$$

and the superscript r denotes the real part and superscript i the imaginary part. Experiments performed by the author indicate that these invariants are no more robust than the $\{|Q_k|\}$. Appendix D shows how to obtain the Fourier coefficients for a discrete image.

Arbter discusses the robustness of the Fourier descriptors to perturbations of the boundary points in his PhD thesis [80] and finds that the perturbations result in errors in the values of the coordinates as well as errors in the parameterization t, with the latter being much more significant. He also finds that the parameterization error can be reduced if one can use significant points along the boundary rather than all the boundary points. Finally, he concludes that the parameterization will be much less sensitive to noise if one uses the signed area rather than the absolute area in (4.17) i.e.

$$t = \frac{1}{2} \int_0^S \{x(s)\dot{y}(s) - y(s)\dot{x}(s)\} \, ds, \tag{4.18}$$

because positive and negative perturbations can partly cancel one another out. This is borne out by the experiment described in section 4.5.1: the signed parameterization of equation (4.18) results in considerably more robust invariants than the unsigned one of equation (4.17).

Fenske & Burkhardt [17] have shown how to generalize the above to grey level images by changing the parameterization so that it is no longer the swept area but the swept intensity which is used. They obtain reasonable performance, but their comparison with moment invariants is flawed: the alleged moment invariants are *not* invariant to affine transformations, nor are they invariant to changes in contrast.

4.4.2   Differential (local) methods

The invariant features we have looked at so far have incorporated information from the whole object or the whole boundary, and hence are termed global invariants. Weiss [6] showed that one can also compute an invariant at any single point on the boundary, using derivatives of curvature (*differential invariants*). As we will see in chapter 6, local invariants are important if one wants to recognize partially occluded objects, since some of the information used by global invariants will be hidden. Differential invariants require a sufficient number of derivatives to remove all the transformation's degrees of freedom, and are hence noise sensitive. Van Gool *et al.* [83] introduced *semi-differential invariants* to reduce the order of derivatives by including some global

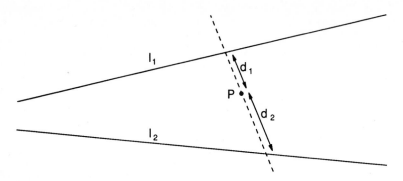

Figure 4.1: An invariant of two lines, a point and its tangent.

The ratio of distances d_1/d_2 is invariant to affine transformations. The solid lines l_1 and l_2 are the lines extracted from the image, P is a reference point on the object boundary and the dashed line represents the tangent at P.

information, such as the coordinates of some reference points (see chapter 1), an idea which has also been adopted by Brill, Barrett *et al.* [84, 85].

The above authors consider invariance to planar projection, and their work will be summarized in chapter 5. Bruckstein & Netravali [22] have used Weiss's approach to obtain differential invariants for affine transformations; unfortunately, they require fifth order derivatives of curvature which makes them too sensitive to noise to be of any practical use. Semi-differential invariants provide a means of reducing the order of derivatives; since affine transformations are a subset of projective ones, one could use the projective semi-differential invariants of chapter 5. Alternatively, one could use semi-differential invariants specifically formulated for affine transformations as follows. An affine transformation has six parameters. A point and a line each have two degrees of freedom, while a tangent at a point has one degree of freedom. Hence we would expect a boundary point and its tangent plus two lines and/or points to give an invariant, and this is indeed so. Figure 4.1 demonstrates the principle for the case of two lines, one point and a tangent at the point.

Lines and points extracted from the boundary are local features; they are also the simplest form of differential features, since they typically depend on the curvature (tangents) of the boundary. It follows that the most robust semi-differential invariants will be those based on lines and points alone, without extracting any further derivatives; the disadvantage is that invariants based on points alone barely take the shape of the boundary between points into account, and hence ignore a great deal of potentially useful information. Point-based invariants are discussed in detail in the following.

A number of authors have considered recognizing objects using the relative positions of point features [78, 21, 19, 86, 7, 87], rather than using the whole image. ... 1 showed how to extract points from a curve; in some cases, points are natu-... s in the first place: a potential application is in recognising fingerprints — ... e locations of points where ridges fork or where they terminate are the main features of a fingerprint; although there is likely to be some nonlinear distortion when taking fingerprints, features invariant to affine transformations are likely to cluster better than other features.

Since an affine transformation is defined by six parameters, any three points can be mapped onto any other three points using such a transformation, and four points provide two independent non-trivial invariants.

Below the theory is expounded; the results of experiments performed to ascertain the stability of the point-based invariants are presented in section 4.5.

Moment invariants of points

One can compute moment invariants based on point features as shown in appendix B; they differ from the moment invariants by not having a factor $|\Delta|^k$ (see equation (4.1)).

To obtain absolute invariants, we can no longer divide by μ_{00}, since this is a constant equal to the number of points. Instead, one can use the fact that area is a relative invariant of weight 1: section B.1 shows that the area of a triangle in centred coordinates defined by the origin and the two points (x_1, y_1), (x_2, y_2) is a relative invariant of weight 1. Hence one could either use the area of the convex hull of the points to replace μ_{00} or, more simply, one can use the symmetric function

$$\mu = \sum_i \sum_{j>i} |x_i y_j - x_j y_i|$$

which satisfies $\mu = |\Delta| \hat{\mu}$ under linear transformations. Clearly, one could alternatively use μ_2:

$$\mu_2 = \sum_i \sum_{j>i} (x_i y_j - x_j y_i)^2.$$

The simplest absolute invariant is given by Q/μ_2, where

$$
\begin{aligned}
Q = \mu_{20}\mu_{02} - \mu_{11}^2 &= \sum_i x_i^2 \sum_j y_j^2 - \left(\sum_i x_i y_i\right)^2 \\
&= \sum_i \sum_j x_i^2 y_j^2 - x_i y_i x_j y_j \\
&= \sum_i \sum_j x_i y_j (x_i y_j - x_j y_i).
\end{aligned}
$$

Note that since Q/μ_2 is symmetric in the $\{x_i, y_i\}$, it is independent of the ordering of the points, a property which will be discussed further in chapter 6.

Invariant affine coordinates of points

Lamdan *et al.* [21] and Hummel & Wolfson [88] appear to have simultaneously introduced the notion of affine coordinates as a means of computing an invariant of a point $\mathbf{v}$ given three other non-collinear points $\mathbf{a}$, $\mathbf{b}$ and $\mathbf{c}$, where $\mathbf{a}^T = [a_x \ a_y]$ etc. The affine coordinates (ξ, η) of $\mathbf{v}$ in terms of the *basis* $\mathbf{a}$, $\mathbf{b}$ and $\mathbf{c}$ is defined so that

$$\mathbf{v} = \xi(\mathbf{a} - \mathbf{c}) + \eta(\mathbf{b} - \mathbf{c}) + \mathbf{c}.$$

Costa *et al.* [86] investigate the properties of the affine coordinates in noise, and show that the coordinates ξ and η are simply ratios of signed areas of triangles. If (**abc**) denotes the signed area of the triangle defined by the three points **a**, **b**, and **c**:

$$(\mathbf{abc}) = (a_x - c_x)(b_y - c_y) - (b_x - c_x)(a_y - c_y)$$

then

$$\xi = \frac{(\mathbf{vbc})}{(\mathbf{abc})}, \quad \text{and} \quad \eta = \frac{(\mathbf{avc})}{(\mathbf{abc})}$$

Note that, in contrast to the moment-based invariants, the affine coordinates depend on the ordering of the four points. One can, of course, obtain invariance to the ordering by forming a permutation invariant of the 24 different possible values; for example, any symmetric function of the 24 values, such as their sum or their product, is invariant to permutations. If some of the 24 values are highly unstable, the resulting symmetric function will also be unstable. This can be circumvented by using the median of the 24 values as the invariant, since unstable invariants are usually either very large or very small (i.e. one of the triangles has a small area).

Signatures

Weiss's differential invariants [6] result in *signatures* of the original curve; a signature is a curve which is a function of the image curve to be recognized but which is invariant to all specified transformations of the image curve (in this chapter affine transformations, in chapter 5 projective transformations), and is generated by plotting one absolute invariant against another, evaluated at all points along the curve. Signatures are not limited to differential invariants: in the affine case, if one has three reference points on the boundary, one can obtain a signature by computing the two independent invariants based on the three reference points plus a point on the boundary, and plotting one against the other as the latter point moves along the boundary. Lamdan *et al.* [21] show how to obtain three reference points given two points on the boundary: the point on the curve furthest from the line joining the two points is invariant under affine transformations and provides the third point needed for a signature.

If one would like to compute invariants based on the boundary section rather than have a continuous signature function, one can modify the above as follows (see figure 6.3 in chapter 6): use the point midway on the line joining the two reference points as the origin; place an odd number n of points at equal spacings along the boundary, using the affine invariant distance metric (equation (4.17)); use point number $(n+1)/2$ to define a third reference point, and then generate invariants using the affine coordinates of the other points relative to the three reference points. (N.B. This technique is the basis for one of the back-projection methods discussed in chapter 6.)

4.4.3   Correlations

To the author's knowledge, the following is a unique description of how to obtain invariance to affine transformations using correlations. Two different approaches are discussed, both of which generalize Perantonis & Lisboa's work [52] discussed in section 2.11.

The first method uses the invariants of four points instead of the angles of a triangle; normalizing the output vector to have unity Euclidean length would take care of the scale factor as before. The main drawback with using the invariants of four points is the computational complexity: for an $N \times N$ image, it is now $O(N^8)$.

The second approach allows one to obtain a number of invariants based on third order correlations: the simplest involves computing the ratio of the area defined by a triplet of points to the area of the whole object. An alternative is to place the origin of the coordinate system at the object's centroid, compute the moment invariant $Q = m_{20}m_{02} - m_{11}^2$ for each triplet of points and form the absolute invariant Q/A^2, where A is the area of the whole object. (Note that both these invariants will not be able to discriminate between reflections).

As discussed in section 2.12.1, a further reduction in computational complexity can be achieved by using equispaced points on the object's boundary and hoping that the lack of a fixed starting point will not make too much difference. We saw in the previous section that the boundary length is not preserved under affine transformations, so one must use an affine invariant distance metric to place the points on the boundary. Arbter *et al.*'s swept-area metric is ideally suited to the task; experimental results are presented later in the chapter (section 4.5.2), and indicate that the ratio of areas is the better of the above two methods.

4.5 Experimental results

Six experiments are described below. The first two compare moment invariants with alternative techniques: first Fourier descriptors, then correlations. In both cases the moment invariants are seen to perform better. Next we will look at how well moment invariants perform on a grey-level image under different lighting conditions and from different viewing angles; the results are very encouraging. The fourth experiment tests the ability of moment invariants to recognize rotational and/or reflectional symmetries; they will be seen to perform well on the former but very poorly on the latter. The penultimate experiment investigates the stability of various point-based invariants as one of the points is perturbed, and the last experiment investigates the variation of the moment invariants for perspective views of non-planar objects (a hammer, a spanner, a file and a chuck-key) — the variation is found to be quite small.

4.5.1 Moment invariants vs Fourier descriptors

In the literature, Dudani *et al.* [2] showed that the rotation-invariant moment features can succesfully discriminate between aircraft silhouettes when combined with a modified nearest-neighbour classifier, so it is very likely that the affine invariant moment features will perform even better. Reeves *et al.* [4] compare moment-based invariants with those based on Fourier descriptors using a minimum Euclidean distance classifier working on a variance-normalized feature space. However, the only affine invariant features they use are the moments of the normalized image (see section 4.3); the other features tested are the scale and rotation invariant moment features and the scale and rotation invariant Fourier descriptors. As one would expect, the affine invariant features performed best (with 93% correctly classified). Considering that the rotation invariant moment features performed far worse than in the experiment reported by Dudani *et al.* [2] (with only around 66% correctly classified, whereas Dudani *et al.*'s classifier performed significantly better than three technical observers

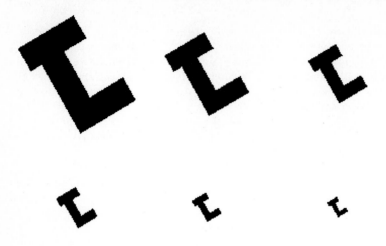

Figure 4.2: The letter τ rotated by 36° over scales 1 to 6.

who were allowed to compare the six models with the images to be classified), it is likely that the affine invariant features would give near 100% accuracy using the modified nearest neighbour classifier instead of the Euclidean distance classifier. Arbter *et al.* [3] have performed experiments on aircraft silhouettes using the affine invariant Fourier descriptors, showing that these features cluster better than the rotation invariant Fourier descriptors, and Cyganski & Orr [58] have performed experiments using template matching of the normalized images, but they do not compare classification accuracy with any other schemes.

The experiments performed test the moment-based features and Arbter *et al.*'s Fourier descriptors on the binary letters 'c' and 'τ' in figures 4.3 and 4.4 over a number of scales, shears and orientations. These letters were chosen because of their similarity to one another; in particular, they both have the same area. The letters were largest for scale 1 and smallest for scale 6 (see figure 4.2); the shapes at scale $i+1$ had half the area of those at scale i, and the initial size was chosen so that all the rotated versions of the original shape would fit on a 128×128 sampling grid at scale 1. Each shape underwent transformations T given by

$$T = 2^{-k/2} \begin{bmatrix} \cos\theta & -\sin\theta \\ \sin\theta & \cos\theta \end{bmatrix} \begin{bmatrix} a & b \\ 0 & 1/a \end{bmatrix}$$

with k taking on values between 0 and 5, $a \in \{1, 2\}$, $b \in \{-\frac{3}{2}, -1, -\frac{1}{2}, 0, \frac{1}{2}, 1, \frac{3}{2}\}$ and $\theta \in \{0, 36^o, 72^o, \ldots, 324^o\}$, giving a total of 140 potential views at each scale. Only those test shapes that fit on the 128×128 grid were used; this means that there were only 14 versions of each letter used at scale 1, 56 versions at scale 2, 120 at scale 3 and the full quota of 140 versions at scales 4, 5 and 6. Appendix D gives the exact size of the letters, as well as details on computing the Fourier descriptors; the boundary tracing algorithm is described and the equations defining the invariant Fourier descriptors are reproduced from Arbter *et al.*'s paper [3].

Figure 4.3: The letter 'c' at scale 5 over all shears and half the rotations.

Figure 4.4: The letter 'τ' at scale 5 over all shears and half the rotations.

Letter	Scales:	R		Γ_8		Γ_5^{-1}	
		1–5	1–6	1–5	1–6	1–5	1–6
c:	Max.	166.7	198.9	52.2	55.4	163.1	162.7
	Min.	30.2	12.0	29.3	27.6	60.0	60.0
τ:	Max.	29.3	50.6	149.2	189.6	48.3	48.3
	Min.	5.53	3.3	72.8	-49.6	13.2	13.2

Letter	Scales:	Γ_7^{-1}		ν_{12}	
		1–5	1–6	1–5	1–6
c:	Max.	164.0	180.7	114.0	119.7
	Min.	58.8	56.1	87.6	86.6
τ:	Max.	50.4	62.4	82.9	94.2
	Min.	13.8	2.7	59.4	30.6

Table 4.2: Experimental results

Features that were able to separate the two letters 'c' and 'τ' over 5 scales. R is a moment invariant (not contrast invariant); Γ_8, Γ_5^{-1} and Γ_7^{-1} are contrast invariants using moment invariants, and ν_{12} is a contrast invariant normalized moment.

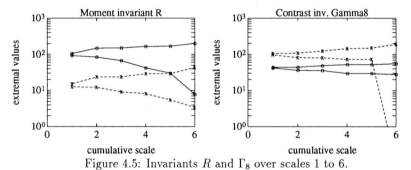

Figure 4.5: Invariants R and Γ_8 over scales 1 to 6.

Dotted line corresponds to 'τ', full line to 'c'.

Table 4.2 shows the performance of the moment-based features that were able to discriminate between the two shapes over scales one to five. The maximum and minimum values obtained when classifying over scales one to five and over scales one to six are listed[1]; despite the images being binary, the contrast invariants performed best, coming very close to separating the letters over all six scales. Figures 4.5, 4.6 and 4.7 show graphically how the features vary more as the range of scales is increased. The circles and crosses indicate the data points; the lower value of Γ_8 for τ at scale 6 was changed from -49.6 to 0.1 to allow plotting with a logarithmic scale.

Figure 4.8 shows the performance of the best two of the first twenty Fourier descriptors using the unsigned parameterization (4.17) [67, 89] and figure 4.9 shows the best two Fourier descriptors using the signed parameterization (4.18). The signed parameterization clearly results in much better performance; nevertheless, the best two features cannot discriminate between the two letters over scales 1 to 5.

---

[1]The values of the features were normalized so that the larger of the two mean values was set to 100.

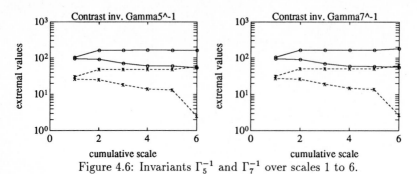

Figure 4.6: Invariants Γ_5^{-1} and Γ_7^{-1} over scales 1 to 6.

Dotted line corresponds to 'τ', full line to 'c'.

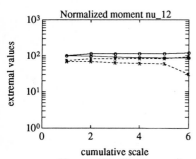

Figure 4.7: The normalized moment ν_{12} over scales 1 to 6.

Dotted line corresponds to 'τ', full line to 'c'.

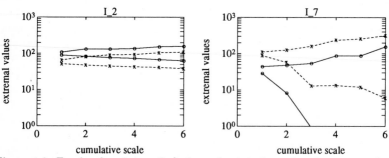

Figure 4.8: Fourier descriptors I_2 & I_7, scales 1 to 6, unsigned parameterization.

Dotted line corresponds to 'τ', full line to 'c'.

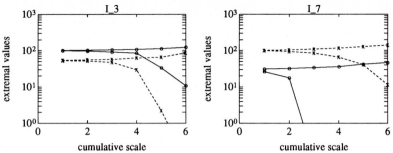

Figure 4.9: Fourier descriptors I_3 & I_7, scales 1 to 6, signed parameterization.

Dotted line corresponds to 'τ', full line to 'c'.

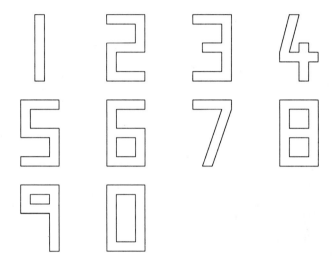

Figure 4.10: The ten digits used in the experiments

As can be seen from figures 4.3 and 4.4, the task of separating the two shapes at scale 5 is not easy, so it is a measure of the discriminatory power of the moment invariants that they perform so well.

4.5.2   Moment invariants vs correlation invariants

The experiments performed below were devised with two purposes in mind: first, to compare the ability of the new correlation invariants with that of the moment invariants at discriminating between affinedly transformed versions of the ten digits '0', '1',..., '9' in figure 4.10; and second, to compare the performance of the VCM as a classifier with that of the nearest neighbour (NN) classifier.

| Invariants: | Ψ_1 | | $\Psi_3 \times 10^{-3}$ | |
Digit	Min.	Max.	Min.	Max.
0	0.0286	0.0318	-0.0003	0
1	0.0053	0.0070	0	0
2	0.0292	0.0329	-0.0002	0
3	0.0208	0.0286	0.0455	0.0766
4	0.0121	0.0172	-0.0514	-0.0364
5	0.0292	0.0329	-0.0003	0
6	0.0220	0.0248	-0.0028	-0.0004
7	0.0235	0.0333	-0.9831	-0.7794
8	0.0195	0.0207	-0.0001	0
9	-0.0006	0.0107	-0.3874	-0.2584

Table 4.3: The moment invariants Ψ_1 and Ψ_3.

The minimum and maximum values of the moment invariants based on Ψ_1 and Ψ_3 over all views of each digit.

Training and test data

An adaptive classifier should be able to classify a wide range of inputs once it has been trained on a limited number of examples. To test this, a data set of 140 views of each digit at each of four scales was generated, and then partitioned into a set of training data and a set of test data. The training data consisted of 16 views of each digit, 8 at scale 1 and 8 at scale 4; figures 4.11 and 4.12 show the 16 training views of the digit '4' (only the pixels along the boundary are plotted). The transformations T used on the digits are the same as those described in section 4.5.1 except that only the first four scales were used. A total of 560 views were obtained for each digit this time, because the 4×4 squares surrounding each view of the digit '4' in figures 4.11 and 4.12 were sampled on a 256×256 grid. It should be mentioned that the experiments test the invariants over a much wider range of scales than the experiments of Perantonis & Lisboa [52], who use scaling factors chosen from a uniform random distribution over the range [0.7 1.3], corresponding to k over the range [−0.76 1.0] — i.e. a span of 1.76 as opposed to 3 in the above case, and far fewer examples at the extremes of scale because of the random distribution.

Image features: moments and correlations

Moment invariants   Since the digits '0', '1', '2', '5' and '8' exhibit two-fold rotational symmetry, moment invariants that contain odd-order moments will be zero when classifying these digits. Hence only one such invariant, ψ_3 in equation (4.6) was used. The remaining invariants were those containing only even-order moments, viz. ψ_1, ψ_5, ψ_6, ψ_7 and ψ_8 in equation (4.6), giving a total of six features. The minimum and maximum values of ψ_1 and ψ_3 are listed in table 4.5.2.

Correlation invariants   Two correlation invariants were tested, one using the ratio of areas and the other using Q/A^2 as discussed in section 4.4.3. In each case the correlations of 50 equispaced points on the object's boundary were computed; to ensure that the points were equispaced, the boundary was interpolated by 'joining the

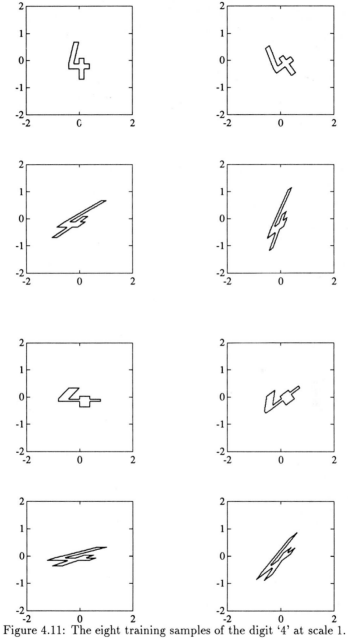

Figure 4.11: The eight training samples of the digit '4' at scale 1.

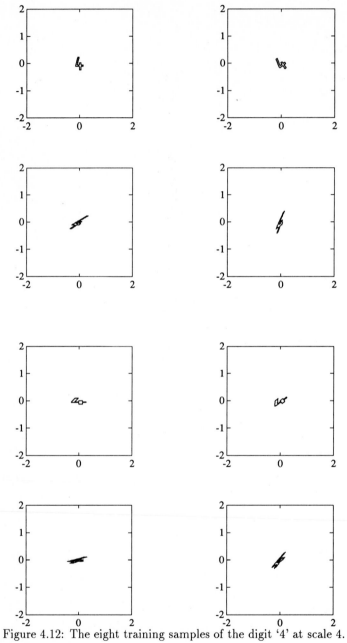

Figure 4.12: The eight training samples of the digit '4' at scale 4.

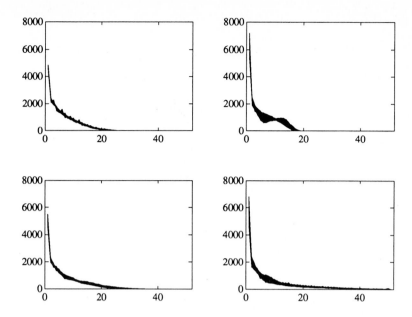

Figure 4.13: The correlation invariants for four digits

The plots show the spread of the 50 correlation invariants based on the area ratio over all 140 views at scale 1 and all 140 views at scale 4, for the digits '0', '1', '7' and '4', clockwise from the top left. The y-axis indicates the number of entries in each bin referred to by the integer values along the x-axis. The fifty points in each plot corresponding to one view are linearly interpolated to emphasize the spread.

dots' — hence the equispaced points were not limited to positions on the sampling lattice. The invariants were partitioned into 50 equal intervals; in the case of the area-ratio, the minimum value was zero and the upper limit was set to 1.5 — any value of the ratio that was larger than 1.5 was allocated to bin 50 (this only occurred with the digit '7'). Figure 4.13 shows the spread of the values in each bin over all views at scales 1 and 4 for the digits '0', '1', '4' and '7' — note the accumulation at bin 50 for digit '7'. Each view of a digit produces a count in each of the 50 bins; the figures were generated by plotting the value in each bin against the bin number and joining the dots with straight line segments, for all 280 views of each digit. It is not immediately apparent that the features discriminate well, but one of the classifiers discussed below achieved good results.

Classifiers: VCM and nearest-neighbour (NN)

The VCM classifier The VCM is an example of a nonlinear system that arises from a simple polynomial expansion of the input variables (see figure 4.14). There has been a lot of work done on the use of the VCM to model nonlinear processes, as the recent tutorial on *Adaptive Polynomial Filters* in the IEEE Signal Processing magazine [90] testifies; however, much less seems to have been done on using the

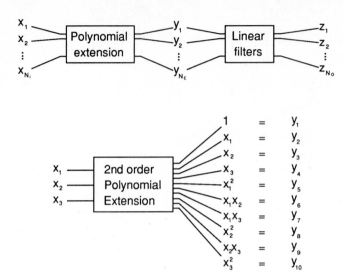

Figure 4.14: The Volterra Connectionist Model (VCM).

The Volterra Connectionist Model (VCM) uses a linear weight vector on a polynomial extension of the input space. The bottom figure shows a second order extension for a VCM with 3 inputs.

VCM to discriminate between different inputs, the main references on polynomial classifiers being those based on Specht's polynomial approximations of the underlying probability density functions [91, 92], those based on Ivakhnenko's Group Method of Data Handling (GMDH) [93, 94, 95] and those based on extensions of linear filters, such as the VCM [96, 97, 70, 98, 99, 100, 101]. The author has shown that a VCM is capable of learning any invariant functions that it can represent, and has verified this experimentally; see reference [99] for more details.

The polynomial extension of the input space used by the VCM is the Kolmogorov-Gabor polynomial, a discrete version of the Volterra expansion for a nonlinear dynamic system [70]. Let $\mathbf{x}$ be the n-dimensional input vector and $\mathbf{y}$ be the kth order extended vector, then the components of $\mathbf{y}$ are all the members of $1, \{x_{i_1}\}, \{x_{i_1} x_{i_2}\}, \{x_{i_1} x_{i_2} x_{i_3}\}, \cdots, \{x_{i_1} x_{i_2} \ldots x_{i_k}\}, 0 \leq i_j \leq n$, where identical terms appear only once (see also Fig. 4.14):

$$\mathbf{x}^T = [x_1 \, x_2 \ldots x_n];$$
$$\mathbf{y}^T = [1 \quad x_1 \, x_2 \ldots x_n \quad x_1^2 \, x_1 x_2 \ldots x_n^2 \cdots x_1^k \ldots x_n^k].$$

The number of terms in a Kth order extension of an N-dimensional input space $T(N, K)$ is

$$T(N, K) = 1 + \sum_{k=1}^{K} \frac{(k + N - 1)!}{k!(N - 1)!}.$$

The output of the VCM given an input $\mathbf{x}$ is given by

$$z = \mathbf{w}^T \mathbf{y}$$

where $\mathbf{y}$ is the extended input vector and $\mathbf{w}$ is the vector containing the weights, which we hope to set to values that allow z to take on non-overlapping values for different input classes. The VCM can easily be extended by using more than one weight vector and hence more than one output, as shown in figure 4.14.

The VCM classifier is a linear classifier acting on a polynomial extension of the input space, as depicted in figure 4.14, and thus needs at least as many training samples as the size of its weight vector. Having chosen to use a total of 160 training examples, we are thus forced to use a VCM with no more than 160 weights. Hence we are limited to a linear (first order) classifier for the 50 correlation features, and to a second order classifier for the 6 moment invariants — the latter requiring 28 inputs. As a result, the VCM was only used to classify the digits based on the moment invariants.

The VCM classifier as discussed in references [98, 99] uses a single weight vector to classify all classes. This will, however, not always be succesful at discrimination, especially when the order of the nonlinearity is limited as in the present case. The reason is that, by using a single weight vector for all classes, one is forcing the classifier to use a discrimination boundary of the same shape to discriminate between all the classes in the input space. As a simple example, take a VCM with two inputs, x_1 and x_2, and consider a second order extension: $\mathbf{e}^T = [1 \quad x_1 \quad x_2 \quad x_1^2 \quad x_1 x_2 \quad x_2^2]$ with a weight vector $\mathbf{w}^T = [0 \quad 0 \quad -1 \quad 1 \quad 0 \quad 0]$. Furthermore, let us assume that we have four classes A, B, C, D, with average outputs -1.5, -0.5, 0.5 and 1.5 respectively. The discrimination boundaries in the output space are simply points, since the output is one-dimensional; they are -1, 0 and 1 (i.e. an output between -1 and 0 is classified as coming from class B, etc.). If we look at the discrimination boundaries in the input space, we see that they have the form $\mathbf{w}^T \mathbf{e} = k$, $k = -1$, 0, 1; in this case we have $-x_2 + x_1^2 = k$, or $x_2 = x_1^2 - k$ — a family of parabolae (see figure 4.15). Hence we see that using a single scalar output for all classes forces the discrimination boundary in the input space to have the same form $\mathbf{w}^T \mathbf{e} = k$. The Rayner mapping is used to maximize the ability of such a constrained classifier to discriminate between classes, with some success [98].

In cases where the single-weight VCM does not perform well, one can use one weight for each class and set the desired output to be 1 for the chosen class and 0 for the others. Given M classes, one needs to train the network M times on the entire training set; however, since the training is that of a linear classifier, it is very swift and the increase in training is not a serious drawback. To classify an unknown input, one computes the output for each weight vector and assigns the input to the class whose weight vector resulted in the largest output.

The nearest-neighbour (NN) classifier The nearest-neighbour classifier has been shown by Duda & Hart [97] to perform well in the large sample case; indeed, they prove that its error-rate is no more than twice that of the optimum Bayes classifier. Analysis of its performance using small training sets is difficult, but numerous experiments indicate that it is generally very good [97, 2, 5, 102]. The NN classifier works by assigning an unknown input to the class of the training example closest to it in the input space. Of course, when using truly invariant features one training example per class would suffice; the errors introduced by sampling the image mean that the invariant features actually vary, requiring more than one training sample per class. The main drawback of the NN classifier is that all the training samples have to be

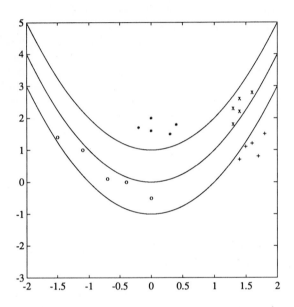

Figure 4.15: Parabolic discrimination boundaries in a 2-D input space.

An example where a two-input single-output second order VCM can discriminate between five classes (examples of inputs from each class are denoted by the stars, crosses, etc.). Note that the discrimination boundaries are parabolae defined by $y = x^2 - k$, $k = \{-1, 0, 1\}$.

stored, and the computational complexity of searching for the nearest neighbour is relatively high.

Questions naturally arise about which distance metric to use and whether to normalize the inputs first. Typically one uses the Euclidean distance, but one must be wary to normalize the inputs so that large fluctuations in one input do not swamp small but possibly significant fluctutions in another input. Below the standard deviation of each input over the training data is normalized to be equal to 1; as we shall see, this works well. The NN classifier was used to classify the digits using the correlation features as well as using the moment features. For the latter, it was also investigated whether the NN classifier performs better on a second order polynomial extension of the input space than on the original input space.

Results

The correlation invariants using Q/A^2 performed significantly worse than using the area ratio, and hence will not be discussed further. Likewise, the single output second order VCM classifier was unable to discriminate adequately, nor was the multiple output first order VCM classifier, so these will also not be discussed further.

Since the digits '2' and '5' are identical bar a reflection, and neither the moment invariants nor the area-ratio correlation invariants can detect reflections, these two digits were assigned the same class during these experiments, giving nine classes in total. If an input is classified as "'2' or '5'", one could use normalization to detect which of the two digits is present.

The classifiers that did perform well were the second order VCM using nine weight vectors, and the NN classifiers using moments, using the second order extension of the moments and using the area-ratio correlation features. In the tables these classifiers are referred to as VCM, NN1, NN2 and NN-C respectively. Table 4.4 shows the performance of the classifiers for each class at each scale as a percentage of the test samples that were correctly classified (N.B. scales 1 and 4 each had 132 test samples for each digit, while scales 2 and 3 each had 140). Table 4.5 shows the performance for each class over all four scales, along with the performance over all scales and all classes. The error rates for the above four classifiers were 1.2%, 1.7%, 2.0% and 3.7% respectively, indicating that the VCM classifier is not only the faster classifier, but also the most reliable one. Of the 4120 test images at scales 1 to 3 it only misclassified 6, and of the 5440 over all four scales it only misclassified 67.

The results indicate that the moment-based features are more robust in this case than the correlation-based features; it would be interesting to know to what extent this is a function of using equispaced points along the boundary rather than using all the points in the image, but at present it is computationally infeasible to perform an experiment using the latter method.

Perhaps surprisingly, the results also indicate that the NN classifier performs slightly worse when using a second order extension as the input rather than the simple inputs, whereas the linear classifier performed very well using the second order extension, but very badly using the simple inputs. It is interesting to note that the VCM was able to correctly classify the digits '0' and '2' & '5' over all four scales and the NN classifiers could not, whereas the latter correctly classified the digits '7' and '9' which the VCM could not. Also, the correlations outperformed the VCM plus moments on digits '6', '7' and '9'.

Input class:		0	1	2, 5	3	4	6	7	8	9
Classifier	Scale									
VCM	1	100	100	100	100	100	100	100	100	100
	2	100	100	100	100	100	100	100	100	100
	3	100	100	100	100	100	97	99	100	100
	4	100	100	100	98	100	89	77	100	89
NN1	1	100	100	100	100	100	100	100	100	100
	2	100	100	100	100	100	99	100	100	100
	3	100	100	90	100	100	88	100	100	100
	4	97	100	92	98	100	81	100	100	100
NN2	1	100	100	100	100	100	100	100	100	100
	2	100	100	100	100	100	99	100	100	100
	3	100	100	90	100	100	84	100	100	100
	4	97	100	92	98	100	79	100	100	100
NN-C	1	95	100	99	88	96	100	100	100	100
	2	91	100	93	84	99	100	100	100	100
	3	89	100	92	81	100	99	100	100	100
	4	86	97	95	93	99	93	100	100	100

Table 4.4: Performance of the four classifiers at each scale.

The entries in the table show the percentage that were correctly classified, rounded to the nearest percent (no value was rounded up to 100).

Input class:	0	1	2, 5	3	4	6	7	8	9	
Classifier										Overall
VCM	100	100	100	99.6	100	96.7	93.9	100	97.4	98.8
NN1	99.3	100	95.8	99.6	100	92.1	100	100	100	98.3
NN2	99.3	100	95.6	99.6	100	90.3	100	100	100	98.0
NN-C	90.6	99.3	94.6	86.6	98.7	98.2	100	100	100	96.3

Table 4.5: Performance of the four classifiers over all scales.

The entries in the table show the percentage that were correctly classified, to one decimal place.

Figure 4.16: Heidelberg postcard, view 1, light.

4.5.3 Moment invariants of grey-level images

Below an experiment is described that computed the value of the standard affine moment invariants for different views of grey level images. The objects to be discriminated were postcards and were viewed from two different positions and under two different lighting conditions, resulting in four views per card. The automatic gain control of the CCD camera was switched off; although care was taken that the brighter of the two images in each case was not washed out, both the lighter and the darker images contain pixels of maximum intensity, implying that the camera was saturated in both cases, and hence that the linear model $f' = cf + b$ does not apply exactly. Nevertheless, the results indicate that the moment invariants provide very good discrimination.

Performance using the contrast invariants on unnormalized images was poor; however, normalizing against brightness and contrast as discussed in section 2.8 resulted in excellent performance. Three distinct postcards were used; they depict views of Heidelberg, Venice and the Scottish highlands, and are shown in figures 4.16 to 4.20.

Figure 4.21 shows graphically the values of the moment invariants for each of the three postcards over all four views; the invariants are numbered as follows[2]: $1 = J_1$; $2 = \psi_1$; $3 = \psi_2$; $4 = \psi_3$; $5 = \psi_4$; $6 = \xi_1$; $7 = \psi_5$; $8 = \psi_6$; $9 = \psi_7$; $10 = \psi_8$; $11 = \xi_2$; $12 = \psi_{10}$ and $13 = \psi_9$. Invariants 1, 7 and 8 provide particularly impressive performance.

Fenske & Burkhardt [17] use Fourier descriptors for grey-level images (postage stamps), but do not obtain 100% discrimination; furthermore the moment functions they compare the Fourier descriptors with are *not* invariant to affine transformations, nor are they invariant to changes in contrast, so the results must be discounted.

---

[2]The object's *shape* is used to compute the centroid, so μ_{10} and μ_{01} based on the *intensity* values will not generally be zero; hence J_1, which is a function of first order moments, can be used.

Figure 4.17: Heidelberg postcard, view 1, dark.

Figure 4.18: Heidelberg postcard, view 2, dark.

Figure 4.19: Venice postcard, view 1, dark.

Figure 4.20: Scottish highlands postcard, view 1, dark.

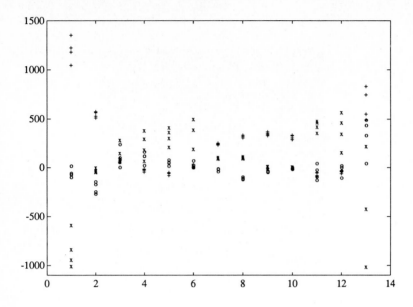

Figure 4.21: Moment invariants for grey-level images.

The values of the invariants plotted against their indices (see text).

4.5.4   Detecting object symmetries

The experiments described below to test how well moment invariants can detect rotational and/or reflectional symmetries were performed on affine transformations of the digits '0' to '9' used earlier and shown in figure 4.10. Note that the digits '0', '1' and '8' possess both reflectional symmetry and two-fold rotational symmetry, and the digits '2' and '5' are reflections of one another and each possess two-fold rotational symmetry. The left column of figures 4.22, 4.23 and 4.24 show four views of the digits '2', '4' and '5' after various affine transformations, and the right column shows the normalized versions. The plots of each digit are obtained by 'joining the dots', so they appear smoother than they would if displayed as a binary image of pixels. The images on the left are sampled on a 128×128 grid (they correspond to scale 3 in the experiments described in chapter 7). Table 4.6 shows the values of the complex moment c_{21} for each of the four views, along with the skew invariant H and the imaginary part of the complex moment n_{40} of the normalized image. In order to use normalization one must be able to detect rotational symmetry; as the table shows, c_{21} is two orders of magnitude smaller for the digits '2' and '5' than for the digit '4', indicating that one can detect rotational symmetry. Nevertheless, one needs to define a threshold for $|c_{21}|$, which is not necessary when using moment invariants.

To discriminate between the digits '2' and '5' the sign of the skew invariants H and $\text{Im}\{n_{40}\}$ must be invariant over different views and be different for each digit. As the table clearly shows, this is not the case for either of the skew invariants, so we must conclude that they are not suited to discriminating between objects that are reflections of one another. Figures 4.22 and 4.24 do however indicate that one should

Digit:		'2'	'5'	'4'
c_{21}	1	151 + 151i	151 + 151i	97200 + 252000i
	2	166 + 166i	166 + 166i	190612 + 184042i
	3	81.3 + 81.3i	81.3 + 81.3i	-9690 + 41700i
	4	80.3 + 80.3i	80.3 + 80.3i	26100 + 62700i
$H \times 10^{-9}$	1	-4.20	4.29	0.186
	2	-4.74	4.82	3.37
	3	0.347	-1.66	0.480
	4	-0.926	0.743	0.755
$\text{Im}\{n_{40}\}$	1	0.0402	-0.0401	0.733
	2	0.116	-0.139	-0.0598
	3	-0.00549	0.0248	0.490
	4	0.0140	-0.0112	0.796

Table 4.6: Detecting symmetries over four views of '2', '4' and '5'.

The values are given to two significant figures. The numbers in the leftmost column refer to the rows in figures 4.22 to 4.24.

be able to use the normalized versions of the digits '2' and '5' to discriminate between them, for example by using template matching. Table 4.3 shows the minimum and maximum values of the moment invariants Ψ_1 and Ψ_3 based on the algebraic invariants Q and I, over all views and all scales used in the experiments described in the previous section. Since I contains odd order moments, we would expect it to be close to zero for the digits '0', '1', '2', '5' and '8' with two-fold rotational symmetry. The results indicate that this is indeed the case.

4.5.5 Stability of invariants of four points

When classifying sets of four points (as is required in chapter 6), one desires the invariants to be stable: a small deviation in the position of one or more of the points should only result in a small deviation in the value of the invariant. To test this for the invariants presented in section 4.4.2, I have taken the three points shown in figure 4.25 and allowed the fourth point to vary on the 20×20 grid of points within the square in the figure. Figure 4.26 shows a contour plot and a mesh plot of the affine coordinates ξ and η, where $\mathbf{a}^T = [0 \ 5]$, $\mathbf{b}^T = [15 \ 0]$, $\mathbf{v}^T = [25 \ 25]$ and $\mathbf{c}$ was the point that varied, while figure 4.27 shows the same for the invariant formed by taking the median of the 24 distinct values of ξ obtainable by permuting the labelling of the four points (the twelfth value was taken as the median). Figures 4.28 and 4.29 show the results for the moment invariants Ψ_1 to Ψ_4.

The results show that the affine coordinates, based on knowledge of the order of the points, are generally stable. Costa *et al.* [86] point out that the affine coordinates can become unstable when the area defined by the three basis points becomes small; this will be discussed further in chapter 6, where schemes for their use are discussed. Introducing invariance to the ordering of the points introduces locally unstable points, as can be seen from figures 4.27 to 4.29; however, the number of unstable regions is small, so in practice the invariants could still be useful.

In addition to stability, one would like the invariants to discriminate between

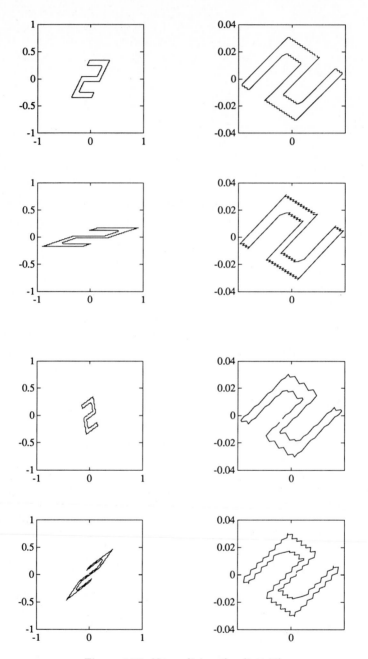

Figure 4.22: Normalizing the digit '2'.

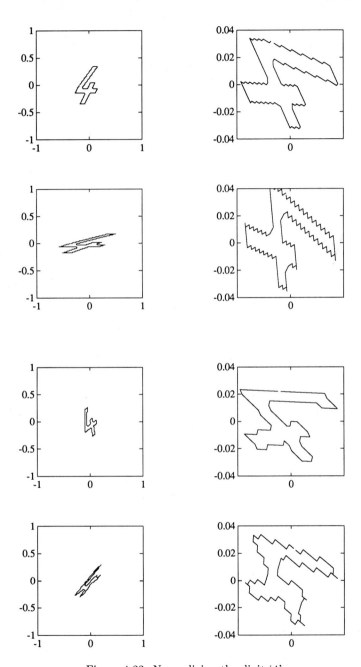

Figure 4.23: Normalizing the digit '4'.

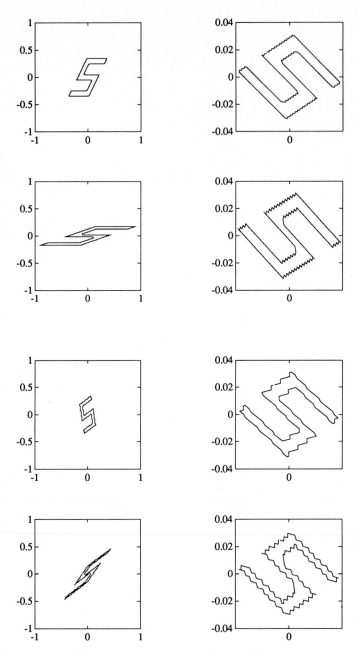

Figure 4.24: Normalizing the digit '5'.

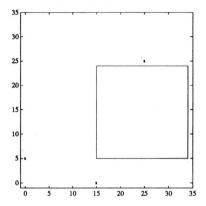

Figure 4.25: The points used to test stability.

The three crosses indicate the three static points; the square denotes the 20×20 region in which the fourth point moved. The axes indicate the pixel numbers.

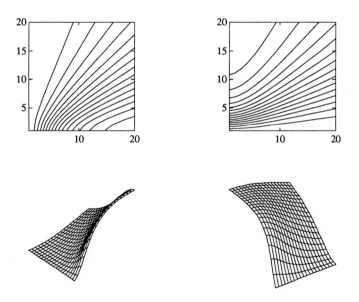

Figure 4.26: The affine coordinates.

Top row: *contour plots.* Bottom row: *mesh plots.* Left column: ξ. Right column: η. *Fifteen levels are shown in these plots.*

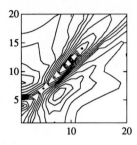

Figure 4.27: The median of affine coordinates.

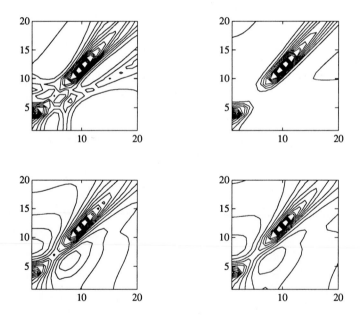

Figure 4.28: Contour plots of the moment invariants.

Clockwise from top left: Ψ_1, Ψ_2, Ψ_3 *and* Ψ_4. *Fifteen levels are shown in these plots.*

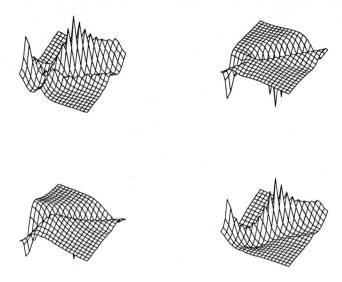

Figure 4.29: Mesh plots of the moment invariants.

Clockwise from top left: Ψ_1, Ψ_2, Ψ_3 *and* Ψ_4.

different quartets; a measure of this is the size of regions in which the invariant is close to a particular value — the smaller such regions, the better the discrimination. The results show that knowing the ordering of the points will allow better discrimination — the invariants shown in figures 4.27 to 4.29 have larger regions of low variation.

4.5.6 Stability of moment invariants under perspective distortion

Figure 4.30 shows a spanner, a hammer, a file and a chuck-key lying on a table viewed from above (0°) and figure 4.31 shows the same objects viewed obliquely, at an angle of approximately 42° to the vertical. The camera was 60cm away from the centre of the image plane; the longest object, the hammer, is 27cm long, which means its depth varied by 18cm. Hence the depth variation over mean distance is equal to 0.3, much larger than the 0.1 rule of thumb used by Thompson & Mundy [11] (see chapter 1). Note that since the objects are not planar, invariants of planar objects are only likely to be useful when the viewing angles are not very large, in which case the effects of perspective should not be as important as when recognizing planar objects from all angles.

Table 4.7 shows the results for the moment invariants that varied least over the two views. The non-contrast invariant features are not as stable as the best contrast invariant, Γ_3, especially for the chuck-key; this could well be caused by errors in estimating the chuck-key's area, since an error of one pixel in the width of the 'arms' will result in a significant change in the overall area estimate. The table shows that most of the invariants are less invariant than one might wish, but they do show

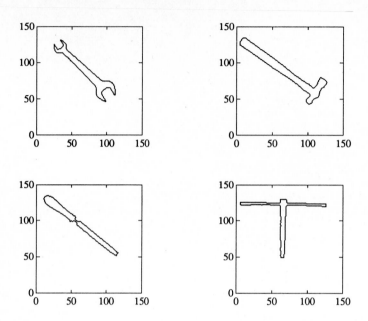

Figure 4.30: The spanner, hammer, file and chuck-key from above.

The axes indicate the pixel numbers.

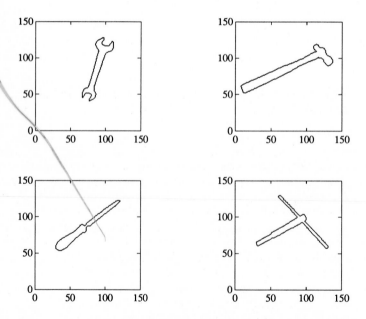

Figure 4.31: The spanner, hammer, file and chuck-key from 42°.

The axes indicate the pixel numbers.

Object:		Spanner	Hammer	File	Chuck-key
Invariant	View				
ψ_1	1	0.020	0.023	0.0090	0.27
	2	0.017	0.017	0.0094	0.15
ψ_5	1	0.0042	0.0070	0.00058	0.77
	2	0.0030	0.0039	0.00062	0.24
ψ_7	1	0.0029	0.0047	0.00049	0.54
	2	0.0020	0.0027	0.00054	0.17
Γ_3	1	-0.1225	-0.1095	-0.1228	-0.1243
	2	-0.1225	-0.1031	-0.1249	-0.1243

Table 4.7: The moment invariants for perspective images.

The values of the moment invariants for the two views of the four objects under perspective. The results are given to two significant figures, except for Γ_3, which is given to four.

that discrimination is possible. All the non-contrast invariant features in the table show at least an order of magnitude difference between the chuck-key, the file and the other two; only the contrast invariant Γ_3 can distinguish between the hammer and the spanner over both views.

Of course, one would have to test the invariants over many more angles to see whether discrimination under this level of perspective is actually achievable; the previous results certainly indicate that it will be possible under approximately orthographic projection.

4.6 Conclusions

Above we have seen a number of different approaches to forming functions invariant to affine image transformations: moment invariants, moment-based normalization, Fourier descriptors, differential (local) invariants and correlation invariants. We saw in chapter 1 that we would like invariant features to index the correct model in a database; normalizing an image followed by template matching appears to be robust to additive noise [36], but it achieves this at the expense of indexing into the database — each model must be matched with the normalized image. The remaining features all provide indexing, so how do their performances compare?

The experiments indicate that, of all the above invariants, the global moment invariants (rather than the point-based ones) perform best, both on binary images and on grey-level images. Nevertheless, in specific cases other invariants may provide better performance — for example, the correlation invariants classified one of the digits better than the moment invariants.

Particular emphasis was placed on the invariants' performance with coarsely sampled images; the moments perform well, and are more robust than the Fourier descriptors based on Arbter's modification [80] to the original affine arc-length parameterization [3] (see section 4.4.1).

The moments were also seen to be reliable at detecting rotational symmetries, but completely unable to detect reflectional symmetry; it appears that normalization followed by template matching is necessary if one wants to distinguish between two shapes that are reflections of one another.

Finally, the permutation-invariant point-based features were seen to be considerably less stable than those that depend on a priori knowledge of the ordering, such as the affine coordinates.

Recognition schemes that make use of the above features to recognize partially occluded objects will be discussed in chapter 6; first, let us look at features invariant to projective transformations.

Chapter 5

Invariance to Projective Transformations

5.1 Introduction

Planar projective transformations are the most general transformations we will consider, and most closely approximate the imaging process. This chapter is devoted to image functions that are invariant to these transformations. The nonlinear nature of projective transformations means that one can no longer use image moments to derive invariants; instead, one must extract image primitives. We will start by looking at local invariants based on points, lines and derivatives of curvature. Some experiments testing the stability of point and line invariants are presented, and a number of semi-differential invariants presented by van Gool *et al.* [103] and by Brill *et al.* [84] are summarized. The next section discusses the use of algebraic invariants of polynomial plane curves to obtain invariant functions. The section focusses on the question of how to fit polynomial curves to image curves, concluding with a summary of some experimental results. Section 5.4 looks at a potentially interesting approach to obtaining invariance to camera rotations and to perspective transformations proposed by Kanatani [104, 13] and Lenz [34] respectively, and discusses why the technique is in fact flawed. Section 5.5 concludes the chapter with a brief summary.

The chapter only considers invariants based on a single view, which is quite sufficient for planar objects. Burns *et al.* [1] show that no single-view invariants are possible for general points in 3D. Invariants do however exist for stereo views from an uncalibrated pair of cameras: Barrett *et al.* [7, 85] have derived a cross-ratio for six general non-coplanar points in 3D; Mohr *et al.* [105] derive an invariant of six points using a different approach, which also allows them to derive invariants for other configurations of lines and points, in particular two pairs of coplanar lines; van Gool *et al.* [106] derive semi-differential signatures for non-planar curves, and Brill *et al.* [84] generalize the methods discussed in section 5.2.3 below to obtain invariants for non-planar curves.

5.2 Differential (local) invariants

As we saw in chapter 4, differential invariants use derivatives of the image boundary's curvature in some form or another, either indirectly when used to extract points and

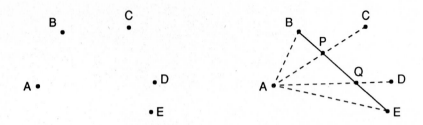

Figure 5.1: Five coplanar points define four collinear points.

The cross-ratio of the points B, P, Q and E is invariant to projections.

lines or directly when used in differential or semi-differential invariants. Below, we will first look at how to use reference points in the image to compute invariant functions, and then investigate their stability to perturbations. Following this, two types of semi-differential invariants are summarized, the signatures of van Gool *et al.* [103] and the two-point invariant of Brill *et al.* [84]. Pure differential invariants, introduced by Weiss [6], require seventh order derivatives of the boundary which makes them too noise sensitive to be of any practical use [107], although Weiss has since found a potentially robust technique that involves approximating part of the curve by a polynomial [108]. Nevertheless, they will not be discussed further.

5.2.1   Invariant functions using reference points alone

Chapter 1 showed us how to extract reference points from an object's boundary using bitangents or maxima of curvature, and chapter 3 showed us that the cross-ratio of four collinear points or five coplanar points is invariant to planar projective transformations. A projective transformation has eight degrees of freedom, so we expect two independent invariants based on five points. Counting degrees of freedom allows one to obtain further invariants based on points and lines: lines and points each have two degrees of freedom and a tangent at a point one — its slope. Hence three points and their tangents or three lines, one point and its tangent (each with nine degrees of freedom) provide an invariant; in both cases constructing the four collinear points needed to form a cross-ratio is straightforward.

One aspect of point-based invariants that has barely been discussed (Forsyth *et al.* [9] mention it briefly) is the fact that the value of the cross-ratio depends on the ordering (labelling) of the points. Below we will first look at an alternative interpretation of the cross-ratio of five points as a cross-ratio of areas, and then go on to discuss means of obtaining invariance to the points' labelling as well as how to maximise the disciminatory power of the five-point cross-ratio. Figure 5.1 shows how one can use five coplanar points to define four points on a line and hence obtain an invariant. Barrett *et al.* [7] show that cross-ratios are easiest to compute using the cross-ratio of the areas of triangles; using the same notation as in section 4.4.2, with A having coordinates **a** etc., allows one to write the cross ratio ρ_A, with A as the vertex common to all triangles, as

$$\rho_A = \frac{(\mathbf{bda})(\mathbf{cea})}{(\mathbf{bea})(\mathbf{cda})}.$$

Clearly the value of the cross-ratio depends on the choice of common vertex and on the choice of labelling for the remaining four points. Although there are 24 distinct labellings of four points, only six distinct values of the cross-ratio arise, three of which are the inverse of the other three [65]. If we write ρ_A as $\rho_1 = \rho_{ABCD}$, then two other distinct values are $\rho_2 = \rho_{CBAD}$ and $\rho_3 = \rho_{DBCA}$, with the remaining three being their inverses. Two examples of symmetric functions that are invariant to the labelling of the four points are

$$
\begin{aligned}
I_{A1} &= \rho_1 + \rho_1^{-1} + \rho_2 + \rho_2^{-1} + \rho_3 + \rho_3^{-1}, \\
I_{A2} &= (\rho_1 + \rho_1^{-1})(\rho_2 + \rho_2^{-1})(\rho_3 + \rho_3^{-1}),
\end{aligned}
$$

where the A in the subscript denotes that vertex A is used as the common vertex. To obtain invariance to choice of vertex point, one can form a symmetric function of the I_{Ai}, such as $I_i = I_{Ai} + I_{Bi} + \ldots + I_{Ei}$, $i = 1, 2$. As it happens, $I_1 \equiv 3$, so one could use I_2, or any other such symmetric function, to obtain invariance to choice of labelling.

An alternative to using symmetric functions is to use the median of the possible values, as discussed in section 4.4.2. In the case of the area cross-ratio, there are a total of 30 distinct values for a set of five points; many of these values will be unstable, because at least one of the areas in one of the denominators is very small. Using the median of the thirty possibilities results in a much more stable invariant than using a symmetric function. Experimental results are presented at the end of this section.

Ideally an invariant function should be reasonably stable and at the same time provide good discrimination between quintets that are not related by a projective transformation. When classifying five points, one can increase the discriminatory power of an invariant by considering the number of points in the quintet's convex hull (see figure 5.2) and by considering the number of collinear points. If four points are collinear, one could form their cross-ratio as an invariant, so this case is ignored in the following, leaving four distinct configurations of five coplanar points to be taken into account: (a) only three points on the convex hull; (b) only four points on the convex hull (N.B. if the inner point is collinear with two pairs of outer points, ρ_A is undefined); (c) all five points on the convex hull; (d) three collinear points. To compute two independent invariants we proceed as follows: in cases (a) and (d), we compute I_{A1} or a suitable median using the two points not on the convex hull or on the line as the common vertex. In case (b), we obtain one invariant by computing I_{A1}, or equivalent, using the point inside the convex hull as the origin. A further invariant can be obtained by using the median of the values obtained by using the four points on the convex hull respectively as the common vertex. Finally, in case (c), which corresponds to figure 5.1, one computes two medians of the thirty possible values, ensuring that one is not the inverse of the other.

Lamdan *et al.* [19] use the affine invariants of four points to obtain reliable recognition even under occlusion; we will see how to generalize their method to use projective invariants in chapter 6.

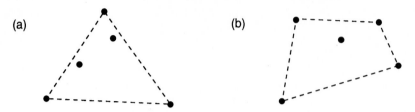

Figure 5.2: Fewer than five points on the convex hull.

The dashed line represents the convex hull. (a) Only three points on the convex hull. (b) Only four points on it.

Stability of the invariant functions

Forsyth *et al.* [9] and Coelho *et al.* [109] have performed some experiments on the stability of the cross-ratio using real images, and conclude that it appears to be very stable. However, the experiments were performed under ideal conditions: the shapes were black polygons on a white background, and the images had a resolution of 640 by 484 pixels. Recently, Sanfeliu *et al.* [110] have analyzed theoretically how imaging errors affect the cross-ratio. Below we will look at the results of an experiment designed to show how uncertainties in the location of a reference point (due to coarse sampling of a distant or small object, and errors in the peak curvature detector) affect the cross-ratio and the median of thirty cross-ratios. The test used is essentially the same as that used to test the stability of the affine invariants in section 4.4.2, but with an extra point added (see figure 5.3). Figure 5.4 shows contour and mesh plots of the cross-ratio ρ_A, where points A, B, D, E have coordinates (0,5), (15,0), (25,25), (15,25) and point C varies over the 20×20 range shown in figure 5.3.

The cross-ratio ρ_A in fact varies fairly uniformly over the regions that look flat in the plots in figure 5.3, this fact being masked by the size of the ridge (plotted as peaks). Hence the experiment essentially backs up Forsyth *et al.*'s findings; however, the results do indicate that the formula for the variance of the the cross-ratio ρ_A given by Forsyth *et al.* [9],

$$\text{Var}\,\{\rho_A\} \propto |\rho_A|$$

is not always helpful when analyzing the stability of the cross-ratio: the cross-ratio of a point near the ridge will have a small magnitude and yet will vary significantly if the point moves towards the peak.

Figure 5.5 shows the stability of the invariant of two points and two lines, where the two lines intersect at the point $(0,5)$ and pass through the points $(15,25)$ and $(15,0)$ respectively (see figure 5.3). The plot shows nearly identical behaviour to that of ρ_A.

As with the affine invariants, introducing permutation invariance decreases the stability of the invariant — more so in this case because the underlying invariant (the cross-ratio) is more complex. The experiments discussed in chapter 6 are designed to test whether one can still use the permutation invariants despite their relative instability.

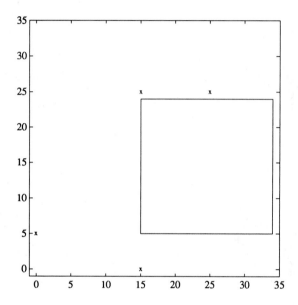

Figure 5.3: The points used to test stability

The four crosses indicate the four static points; the square denotes the 20 × 20 region in which the fifth point moved. The axes indicate the pixel numbers.

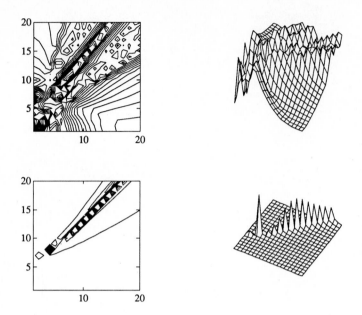

Figure 5.4: Contour and mesh plots of two projective invariants.

Top row: *median of thirty cross-ratios.* Bottom row: *cross ratio ρ_A.* 15 levels are shown in each contour plot. The largest value (∞) but was set to 40 to enable plotting.

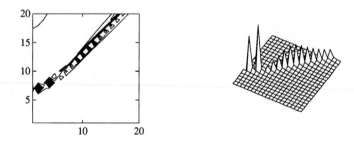

Figure 5.5: Contour and mesh plots of the invariant of two lines and two points.

Fifteen levels are shown in the contour plot. The largest peak value in the figures was in fact ∞, but was set to 40 to enable the plots to be made.

5.2.2 Signatures based on one or two reference points

Van Gool *et al.* [103] use the theory of Lie groups to derive two absolutely invariant signature functions $\tau_1(t)$ and $\tau_2(t)$ that use two reference points $\mathbf{x}_i^T = [x_i \ y_i]$, $i = 1, 2$, and a variable point on the curve $\mathbf{x}^T = [x(t) \ y(t)]$. The first one, $\tau_1(t)$, only uses derivatives at $\mathbf{x}$, but requires them up to second order (below $|\mathbf{a} \ \mathbf{b}|$ is the determinant of the 2×2 matrix with columns $\mathbf{a}$ and $\mathbf{b}$, $\dot{\mathbf{x}}$ and $\ddot{\mathbf{x}}$ are the first two derivatives of $\mathbf{x}$ with respect to t respectively and $\text{abs}\{x\} = |x|$):

$$\tau_1(t) = \int_0^t \text{abs} \left\{ \frac{|\dot{\mathbf{x}} \ \ddot{\mathbf{x}}| \, |\mathbf{x} - \mathbf{x}_1 \ \mathbf{x} - \mathbf{x}_2|}{|\mathbf{x} - \mathbf{x}_1 \ \dot{\mathbf{x}}| \, |\mathbf{x} - \mathbf{x}_2 \ \dot{\mathbf{x}}|} \right\} \, dt. \tag{5.1}$$

The other invariant, $\tau_2(t)$, is potentially more robust to the effects of noise because it only uses first order derivatives:

$$\tau_2(t) = \int_0^t \text{abs} \left\{ \frac{|\mathbf{x} - \mathbf{x}_1 \ \dot{\mathbf{x}}| \, |\mathbf{x}_1 - \mathbf{x}_2 \ \dot{\mathbf{x}}_1|}{|\mathbf{x} - \mathbf{x}_1 \ \mathbf{x} - \mathbf{x}_2| \, |\mathbf{x} - \mathbf{x}_1 \ \dot{\mathbf{x}}_1|} \right\} \, dt. \tag{5.2}$$

Van Gool *et al.* also present the following relative invariants,

$$\tau_3(t) = \int_0^t \text{abs} \left\{ \frac{|\mathbf{x} - \mathbf{x}_1 \ \dot{\mathbf{x}}|}{|\mathbf{x} - \mathbf{x}_1 \ \mathbf{x} - \mathbf{x}_2|^2} \right\} \, dt, \tag{5.3}$$

which requires two reference points but only one tangent (at $\mathbf{x}$), and

$$\tau_4(t) = \int_0^t \text{abs} \left\{ \frac{|\dot{\mathbf{x}} \ \ddot{\mathbf{x}}|^{\frac{2}{3}}}{|\mathbf{x} - \mathbf{x}_1 \ \dot{\mathbf{x}}|} \right\} \, dt, \tag{5.4}$$

which uses second-order derivatives but only requires one reference point. As $\tau_3(t)$ and $\tau_4(t)$ are relative invariants, the signature curves can be arbitrarily scaled along the τ-axis; absolute invariance can be obtained if one can reliably normalize the 'height' of the signature above the t-axis. Alternatively, one could compute scale invariant functions of the signature. In both cases absolute invariance depends on being able to extract the entire signature, for example between a pair of reference points on the curve. If one requires a number of invariant values, rather than a continuous invariant curve (the signature), one can simply compute functions of the signature such as moments etc.

Brown [107] has performed some prelimanary experiments evaluating the stability of the above invariants for discrete curves; as one would expect, they are more robust than pure differential invariants and appear likely to be useful in recognition tasks. However, the experiments did not test for changes in the curve's parameterization.

5.2.3 An absolute invariant based on two reference points

Brill *et al.* [84] derive an absolute invariant based on two reference points, their tangents and their curvatures. This is a remarkable achievement because it only has eight degrees of freedom: tangents and curvatures are scalars (i.e. slope and radius of curvature), so each point with its tangent and curvature has four degrees of freedom.

If one lets $\mathbf{x}_i^T = [x(t_i) \; y(t_i) \; 1]$, $\lambda_i = (a_{31}x(t_i) + a_{32}y(t_i) + 1)^{-1}$ and $\mathbf{A}$ be the matrix with entries $A_{ij} = a_{ij}$, $a_{33} = 1$, then a point $\mathbf{p}_i = [p(s_i) \; q(s_i) \; 1]^T$ is related to $\mathbf{x}_i$ by a projective transformation if

$$\mathbf{p}_i = \lambda_i \mathbf{A} \mathbf{x}_i,$$

where $s_i = s(t_i)$ takes a possible reparameterization of the curve into account. Differentiating twice with respect to t and rearranging gives

$$[\mathbf{p}_i \; \dot{\mathbf{p}}_i \; \ddot{\mathbf{p}}_i] \begin{bmatrix} 1 & 0 & 0 \\ 0 & \dot{s}_i & \ddot{s}_i \\ 0 & 0 & \dot{s}_i^2 \end{bmatrix} = \mathbf{A} \, [\mathbf{x}_i \; \dot{\mathbf{x}}_i \; \ddot{\mathbf{x}}_i] \begin{bmatrix} \lambda_i & \dot{\lambda}_i & \ddot{\lambda}_i \\ 0 & \lambda_i & 2\dot{\lambda}_i \\ 0 & 0 & \lambda_i \end{bmatrix}.$$

The dot means a derivative with respect to t, except $\dot{\mathbf{p}}_i$, which is with respect to s. Taking determinants of both sides gives

$$|\mathbf{p}_i \; \dot{\mathbf{p}}_i \; \ddot{\mathbf{p}}_i| = |\mathbf{A}| \, \dot{s}_i^{-3} \, |\mathbf{x}_i \; \dot{\mathbf{x}}_i \; \ddot{\mathbf{x}}_i| \, \lambda_i^3. \tag{5.5}$$

If one has two points $\mathbf{x}_i$ and $\mathbf{x}_j$ on the curve, then

$$[\mathbf{p}_i \; \dot{\mathbf{p}}_i \; \mathbf{p}_j] \begin{bmatrix} 1 & 0 & 0 \\ 0 & \dot{s}_i & 0 \\ 0 & 0 & 1 \end{bmatrix} = \mathbf{A} \, [\mathbf{x}_i \; \dot{\mathbf{x}}_i \; \mathbf{x}_j] \begin{bmatrix} \lambda_i & \dot{\lambda}_i & 0 \\ 0 & \lambda_i & 0 \\ 0 & 0 & \lambda_j \end{bmatrix}.$$

Taking determinants gives

$$|\mathbf{p}_i \; \dot{\mathbf{p}}_i \; \mathbf{p}_j| = |\mathbf{A}| \, \dot{s}_i^{-1} \, |\mathbf{x}_i \; \dot{\mathbf{x}}_i \; \mathbf{x}_j| \, \lambda_i^2 \lambda_j. \tag{5.6}$$

Using the fact that the curvature k_i at $\mathbf{x}_i$ is given by $|\mathbf{x}_i \; \dot{\mathbf{x}}_i \; \ddot{\mathbf{x}}_i|$, and the perpendicular distance d_{ij} from a point $\mathbf{x}_i$ to the tangent to the curve at $\mathbf{x}_j$ is given by $d_{ij} = |\mathbf{x}_j \; \dot{\mathbf{x}}_j \; \mathbf{x}_i|$, Brill et al. [84] use (5.5) and (5.6) to show that Θ is an absolute invariant under projective transformations:

$$\Theta = \frac{d_{12}}{d_{21}} \left(\frac{k_1}{k_2} \right)^{\frac{1}{3}}.$$

They also show that $\Theta \equiv 1$ for points on a conic.

Note that Θ cannot be used when the two reference points $\mathbf{x}_1$ and $\mathbf{x}_2$ are obtained from the same bitangent, since in that case $d_{12} = d_{21} = 0$. Brill et al. go on to list invariants using more than two points, and also discuss obtaining invariants from stereo views of non-planar curves.

5.3   Global invariants

The only global projective invariants, i.e. those that use information from the whole of the object's boundary without relying on derivatives, known at present are based on the polynomial plane curves. We saw earlier that polynomial plane curves are projected to polynomial curves of the same order; in this section we will see how to use this to obtain image invariants. First we will look at how to extract the coefficients of image curves that are known to be polynomials; then, we will see how this technique can be used to fit polynomials to arbitrary curves to obtain affine invariance. Finally, an alternative approach that involves fitting ellipses to curves in a projectively invariant manner is briefly reviewed [111].

As we will see, the only invariants discussed below that are reasonably stable are those based on pairs of conics, which are of limited use because, at present, one needs one distinct object curve for each conic that is being fitted. As a result, the methods presented below will not be used in chapter 6.

5.3.1 Finding the coefficients and invariants of polynomial object curves

Typically the only polynomial curves that appear in images are lines and conics, which have been investigated by Forsyth *et al.* [112, 9]. A pair of conics have two invariants which are most easily expressed in terms of a matrix of parameters [113, 73]. Since this matrix formulation is unique to conics, we will use a vector formulation in this section. A general kth order polynomial plane curve can be written as

$$\mathbf{a}(k)^T \mathbf{v}(k) = 0, \tag{5.7}$$

where $\mathbf{a}(k)^T = \begin{bmatrix} a_{k,0} & a_{k-1,1} \cdots a_{0,k} & a_{k-1,0} & \cdots & a_{0,k-1} & \cdots & a_{1,0} & a_{0,1} & a_{0,0} \end{bmatrix}$

and $\mathbf{v}(k)^T = \begin{bmatrix} x^k & x^{k-1}y & \cdots & y^k & x^{k-1} & \cdots & y^{k-1} & \cdots & x & y & 1 \end{bmatrix}$,

and (x, y) are the coordinates of points on the curve in the plane. In the following we will drop the parameter k. Hence if a point (x, y) lies on a polynomial curve defined by coefficients $\mathbf{a}$, then $\mathbf{a}^T \mathbf{v} = 0$. This forms the basis for the easiest way of determining the coefficients of a polynomial curve $(x(t), y(t))$, $t \in [0, 1]$, namely by minimizing the algebraic distance [114, 112, 9]. If we let $f_p(x, y; \mathbf{a}) = \mathbf{a}^T \mathbf{v}$, then the algebraic distance is defined as

$$D(\mathbf{a}) = \int_0^1 \{f_p(x(t), y(t); \mathbf{a})\}^2 \, dt.$$

($D(\mathbf{a}) = 0$ if and only if $(x(t), y(t))$ is a polynomial with coefficient vector $\mathbf{a}$.) If $\mathbf{v}(t)$ is the vector $\mathbf{v}$ for a point $(x(t), y(t))$, $D(\mathbf{a})$ becomes

$$\begin{aligned}
D(\mathbf{a}) &= \int_0^1 \{\mathbf{a}^T \mathbf{v}(t)\}^2 \, dt \\
&= \int_0^1 \mathbf{a}^T \mathbf{v}(t) \mathbf{v}(t)^T \mathbf{a} \, dt \\
&= \mathbf{a}^T \left\{ \int_0^1 \mathbf{V}(t) \, dt \right\} \mathbf{a} \\
&= \mathbf{a}^T \mathbf{M} \mathbf{a}, \tag{5.8}
\end{aligned}$$

where $\mathbf{M}$ is a moment matrix (also called a scatter matrix) of the curve, and $\mathbf{V}(t) = \mathbf{v}(t)\mathbf{v}(t)^T$. If we index rows and columns of $\mathbf{M}$ in the same way as entries in $\mathbf{v}$ (i.e. entry $x^p y^q$ has index p, q), we find that each entry $\mathbf{M}_{p_1, q_1; p_2, q_2}$, $p_i + q_i \le k$, is given by

$$\mathbf{M}_{p_1, q_1; p_2, q_2} = \int_0^1 [x(t)]^p \, [y(t)]^q \, dt, \qquad p = p_1 + p_2, \, q = q_1 + q_2,$$

a $(p+q)$th order moment of the curve.

Finding the best fitting polynomial is simply a matter of minimizing (5.8) with some constraint on the coefficients to prevent $\mathbf{a} = 0$. Bookstein [114] mentions a number of constraints that have been used to fit conics to data, in particular $\mathbf{a}^T\mathbf{a} = 1$ (used by Paton [115, 116]) and $a_{0,0} = 1$ (used by Biggerstaff [117], Albano [118] and Cooper & Yalabik [119]). With discrete images the integral in (5.8) is replaced by a summation over all points $\{(x_i, y_i)\}$ on the boundary, so $\mathbf{M} = \sum_i \mathbf{v}_i\mathbf{v}_i^T$, with $\mathbf{v}_i$ being the vector $\mathbf{v}$ at point (x_i, y_i).

Having considered how to find the polynomials' coefficients, we need to ask ourselves whether an absolute projective invariant $I(\mathbf{a})$ of the ternary form with these coefficients will be an absolute invariant under projections of the image[1]. On the face of it a slight problem arises because the coefficients $\mathbf{a}$ can be scaled by an arbitrary factor $k \neq 0$: $\mathbf{a}$ and $k\mathbf{a}$ define identical curves. However, this is of no consequence because, $I(k\mathbf{a}) = I(\mathbf{a})$ for all $k \neq 0$, a property that derives from the fact that $I(\mathbf{a})$ is the ratio of two homogeneous polynomials of the same order in the coefficients (elements of $\mathbf{a}$). In the present section the image curve is assumed to be a polynomial, so if $\mathbf{M}$ is the moment matrix of the original and $\hat{\mathbf{M}}$ is that of the projected version, minimizing the algebraic distance with respect to any of the above constraints results in coefficient vectors $\mathbf{a}_{\min}$ and $\mathbf{b}_{\min}$ that satisfy $\mathbf{a}_{\min}^T\mathbf{M}\mathbf{a}_{\min} = \mathbf{b}_{\min}^T\hat{\mathbf{M}}\mathbf{a}_{\min} = 0$. Hence $I(\mathbf{a}_{\min}) = I(\mathbf{b}_{\min})$.

The above covers invariance to projection. If we limit ourselves to weak perspective (i.e. the affine approximation) and we have a reference point on the image, so that we only need invariance to linear transformations, then one can additionally use the invariants of binary forms. Consider

$$\tilde{a}_{p-l,l} = a_{p-l,l}/\binom{p}{l}, l = 0, 1, \ldots, p,$$

where $a_{p-l,l}$ is an element of the coefficient vector $\mathbf{a}$ in equation (5.7). The $\{\tilde{a}_{p-l,l}\}$ are the coefficients of the pth order binary form $f_p(x, y)$, and can be written as a vector $\tilde{\mathbf{a}}_p$. Since an absolute invariant $J(\tilde{\mathbf{a}}_p)$ of a single binary form satisfies $J(k\tilde{\mathbf{a}}_p) = J(\tilde{\mathbf{a}}_p)$ for all $k \neq 0$, the above theorem implies that an absolute invariant $J(\tilde{\mathbf{a}}_p)$ of the coefficients $\tilde{\mathbf{a}}_p$ obtained as above will be invariant to linear image transformations. One can also use absolute invariants $L(\tilde{\mathbf{a}}_p, \tilde{\mathbf{b}}_q)$, $p \neq q$, of more than one binary form, but one must be careful to ensure that the degree of the numerator and denominator polynomials is the same, so that they are invariant to scaling of the coefficients by k, $k \neq 0$. For example, this procludes the use of J_1/Q (equations (3.4) and (3.6)), since J_1 is third order and Q is second order in the coefficients. Examples of suitable joint invariants are

$$\Pi_1 = \frac{T J_1}{QG} \qquad \text{and} \qquad \Pi_2 = \frac{SQ}{N}. \tag{5.9}$$

(Q, S, T, J_1, G and N are listed in section 3.4.1).

---

[1]Projective invariants are usually written in terms of the coefficients $a_{p-k,k-l,l}$ of the ternary form in (3.9). An invariant $I(\mathbf{a})$ can be obtained from such an expression by setting $a_{p-k,k-l,l} = a_{p-k,k-l}(p-k)!(k-l)!l!/p!$, where $a_{p-k,k-l}$ is one of the terms in $\mathbf{a}$, equation (5.7).

5.3.2 Approximating general curves with polynomial curves

As noted already, other than lines and conics, most image curves are not exactly described by finite order 2-D polynomials. Nevertheless, the above theory can be used to recognize general curves if one can find an invariant way of approximating general curves by polynomial curves [6, 9], so that the invariants of the fitting polynomial remain invariant under transformations of the general image curve.

When the object curve is not a polynomial, the minimum algebraic distance D_{min} is greater than zero, which means that the choice of the constraint (Lagrangian) becomes important. Bookstein [114] considers the problem of fitting a conic to a set of points in an invariant manner, which he defines as follows: if one has a curve C and a transformed version $\hat{C}$, the coefficients of the conic that best fits $\hat{C}$ must be exactly the same as the transformed coefficients of the conic that best fits C. To obtain this he shows that the Lagrangian must be a function that is absolutely invariant to the given transformation (he considered translation, rotation and changes in scale). Forsyth *et al.* [112, 9] point out that one can use a relative invariant as the Lagrangian if one uses the coefficients to compute absolute invariants as discussed above, and if one limits oneself to affine transformations. Since a relative invariant is a polynomial in the coefficients, the minimization becomes tractable, though it is still quite involved [120]. If we ignore translations (e.g. by assuming that we have a known reference point in the image) we need only consider linear transformations, for which $a_{0,0}$ in (5.7) is an absolute invariant. Hence we can use $a_{0,0} = 1$ as the constraint, which makes computing the minimum very simple. This is proved below. (Note that this constraint prevents the curve from passing through the origin).

Theorem If $I(\mathbf{b})$ is an absolute invariant of the kth order ternary form with coefficients $\mathbf{b}$ under linear transformations and we have a set of image points (x_i, y_i), $i = 1, \ldots, N$, that are a linear transformation of a set of object points, then $I(\mathbf{a}_{min})$ will be invariant under all such transformations if

$$\mathbf{a}_{min} = \min_{\mathbf{a}} \left\{ \mathbf{a}^T \mathbf{M} \mathbf{a} \mid a_{0,0} = 1 \right\}.$$

Proof If we write $\mathbf{v}_i^T = [\, \mathbf{w}_i^T \; 1\,]$ and $\mathbf{a}^T = [\, \mathbf{c}^T \; a_{0,0}\,]$, we see that $\mathbf{v}_i$ at a given point is related to $\hat{\mathbf{v}}_i = [\, \hat{\mathbf{w}}_i^T \; 1\,]$ at a linearly transformed point by $\mathbf{v}_i^T = [\, \mathbf{T}\hat{\mathbf{w}}_i^T \; 1\,]$, where $\mathbf{T}$ is a square, non-singular matrix. With $a_{0,0} = 1$ the algebraic distance $D(\mathbf{a})$ becomes

$$
\begin{aligned}
D(\mathbf{a}) &= F(\mathbf{c}) = \sum_i \left(\mathbf{c}^T \mathbf{w}_i + 1 \right)^2 = \sum_i \left(\mathbf{c}^T \mathbf{T} \hat{\mathbf{w}}_i + 1 \right)^2 \\
&= \sum_i \left(\hat{\mathbf{c}}^T \hat{\mathbf{w}}_i + 1 \right)^2 = \hat{F}(\hat{\mathbf{c}}) = \hat{D}(\hat{\mathbf{a}}) \qquad \text{where} \qquad \hat{\mathbf{c}} = \mathbf{T}^T \mathbf{c}.
\end{aligned}
$$

We seek to prove that $I(\hat{\mathbf{a}}_{min}) = I(\mathbf{a}_{min})$. Now, by the definition of $I(\mathbf{a})$,

$$\hat{\mathbf{c}}_{min} = \mathbf{T}^T \mathbf{c}_{min} \quad \Rightarrow \quad I(\hat{\mathbf{a}}_{min}) = I(\mathbf{a}_{min}).$$

Hence we need to prove that $\hat{\mathbf{c}}_{min} = \mathbf{T}^T \mathbf{c}_{min}$. We prove this by contradiction: if it were not true, i.e. $\hat{\mathbf{c}}_m = \mathbf{T}^T \mathbf{c}_{min} \neq \hat{\mathbf{c}}_{min}$, so $\hat{F}(\hat{\mathbf{c}}_{min}) < \hat{F}(\hat{\mathbf{c}}_m)$, we can define $\mathbf{c}_m = (\mathbf{T}^T)^{-1} \hat{\mathbf{c}}_{min}$. From the above,

$$F(\mathbf{c}_m) = \hat{F}(\hat{\mathbf{c}}_{min}) < \hat{F}(\hat{\mathbf{c}}_m) = F(\mathbf{c}_{min})$$

which implies that $F(\mathbf{c}_m) < F(\mathbf{c}_{\min})$ and hence $D(\mathbf{a}_m) < D(\mathbf{a}_{\min})$ which contradicts the definition of $\mathbf{a}_{\min}$. Hence $I(\hat{\mathbf{a}}_{\min}) = I(\mathbf{a}_{\min})$. Q.E.D.

The above shows how to obtain invariant features for general polynomial curves in theory; are these methods robust to the effects of imaging and segmentation errors in practice? The results of Forsyth *et al.* [9] indicate that fitting conics to discrete images of unoccluded objects is reliable. The disadvantage with using the invariants of a pair of conics is that one must fit two coplanar conics to the object contours; to obtain invariants based on a single object contour one must use cubic or higher order polynomials. Below the results of a number of computer simulations performed by the author using the fitting method of the above theorem are summarized; more details can be found in reference [121].

First, it was verified that the invariant Ψ of the quartic ternary form and the two invariants Π_1 and Π_2 were absolutely invariant for points sampled to double precision accuracy; next, the robustness of the invariants of a pair of conics fitted to discrete images of closed curves was investigated and found to be good — this backs up Forsyth *et al.*'s results [112]. Finally, the robustness of the invariants Ψ, Π_1 and Π_2 based on fitting a quartic to a single closed curve was investigated; unfortunately, they are so unstable as to be useless. This was true even when the object curve to be sampled was in fact a quartic polynomial!

5.3.3   Conclusions

The invariants based on minimizing algebraic distance are only robust when fitting pairs of conics to two image boundaries; invariance to affine transformations is obtained using a relative invariant as a Lagrangian [9], but is computationally difficult [120]. Invariance to linear transformations is computationally straightforward using the $a_{0,0} = 1$ constraint. The technique would be more useful if one could reliably use the invariants of a quartic approximation to a single image curve, but the above results indicate that this is not feasible; Zisserman & Rothwell [122] have obtained similar results using the relative invariant as the Lagrangian.

The main drawback with the use of invariants of polynomial plane curves as discussed above is the lack of a robust method of approximating a general object curve with a polynomial curve. Recently, Carlsson [111] has put forward an alternative curve-fitting method that is invariant to projective transformations. He finds a unique and small number of ellipses that inscribe the object boundary with four contact points; the invariants of pairs of these ellipses could be used to characterize the object's shape. At present the method only applies to polygonal shapes; hopefully it will soon be adapted to cope with general shapes.

5.4   Invariance to camera rotation

Kanatani [104, 13] has investigated a subset of projective invariance, namely invariance to camera rotations. This is potentially useful if one has a camera scanning a room with fixed fittings, and also provides a method of detecting when one of the objects in the room has moved (p.193, [13]), but is otherwise of limited use. Kanatani's book [13] introduces the fundamentals of group theory, but only uses them to derive functions invariant to camera rotations. Chapter 6 of his book discusses how to represent 3D rotation so that the rule of composition becomes simple, and shows that this can be achieved using the Cayley-Klein parameters. He then attempts to show that these parameters are linked to the coordinates obtained by stereo projection, which maps the points on a sphere onto points on a plane as follows: A point (X, Y, Z) on the sphere $X^2 + Y^2 + Z^2 = 1$ is mapped onto the point (x, y) on the xy-plane where the ray connecting the point (X, Y, Z) and the 'south pole' $(0, 0, -1)$ intersects the plane. x and y are related to X, Y, and Z as follows:

$$x = \frac{X}{1+Z}; \qquad y = \frac{Y}{1+Z}.$$

A rotation of the sphere moves a point (X, Y, Z) to a different point (X', Y', Z'), and results in its projection (x, y) moving to (x', y'). The chapter allegedly proves that any 3D rotation is equivalent to the following coordinate transformation in the image plane, where $z = x + iy$, $i = \sqrt{-1}$:

$$z' = \frac{\gamma + \delta z}{\alpha + \beta z}, \qquad \gamma = -\beta^*, \quad \delta = \alpha^*, \quad \alpha\delta - \beta\gamma = 1. \qquad (5.10)$$

The constraints mean that the above equations have three degrees of freedom, as one would expect for a rotation in 3D. However, if a 3D rotation causes the above image coordinate transformation, one could use the algebraic invariant of four points of equation (3.11) to obtain a function invariant to camera rotations: simply replace the roots t_i by the complex coordinates z_i. The resulting invariant S^3/M is invariant to all complex fractional linear transformations of z, of which the transformation in equation (5.10) is a subset. The author has verified that S^3/M is indeed invariant to fractional linear transformations, and that it is *not* invariant to object rotations; hence there would appear to be a serious flaw with Kanatani's use of the Cayley-Klein parameters for image invariance.

Lenz [34] generalizes the above to show that the general complex linear fractional transformation with $\alpha\delta - \beta\gamma = 1$ represents solid motion in 3D, potentially a much more powerful result than Kanatani's, which is limited to camera rotations. However, Lenz's proof appears to be flawed, and certainly the invariant S^3/M of four points is not invariant to solid motion in 3D.

5.5 Conclusions

Above, we saw that one must extract features such as points, lines and conics and/or derivatives of curvature from an image if one wants to compute projectively invariant functions. The most robust features appear to be bitangent lines (see chapter 1); bitangent points contain more information and will generally be stable too, except when they lie on a segment of curve with very low curvature. The invariant functions of lines and/or points presented in chapter 3 depend on the ordering of the points; a number of ways of obtaining invariance to the ordering were presented, and their stability investigated. The permutation invariant features turn out to be considerably less stable than those using labelled points, not surprising when one considers that the latter use more prior information. We also saw that the ability of features based on five points to discriminate between different quintets can be increased by considering the number of points on the convex hull.

In section 5.2 we saw how tangents and curvatures can be used to reduce the number of reference points required to compute an invariant: first, four signatures derived by van Gool *et al.* [103] were presented, two of which require two reference points plus tangents, and second, Brill *et al.*'s [84] invariant based on two points, their tangents and curvatures was discussed.

Section 5.3 discussed how algebraic invariants of ternary forms can be used to obtain invariants based on image curves. Experiments that test whether the invariants of a quartic polynomial curve were robust under linear transformations were summarized; the results show that fitting image curves with cubic or higher polynomials by minimizing the algebraic distance does not result in robust invariant functions, but that fitting pairs of conics is generally robust. It appears therefore that, until one can find a reliable curve-fitting method, one must use van Gool *et al.*'s signatures if one requires global invariants of a single image curve; however, these signatures do require two reference points. Whether they will prove to be stable remains to be seen, although Brown's preliminary results are encouraging [107]. Finally, section 5.4 demonstrated that Kanatani's [104] proposed application of the Cayley-Klein parameters cannot be used to obtain invariant features.

This chapter concludes our journey through the realms of invariant features; we now move on to discuss schemes that use these features to recognize partially occluded objects. Of the invariants presented in this chapter, those based on lines and points as well as the differential invariants will be considered in the next chapter.

Chapter 6

Recognizing Partially Occluded Objects

6.1 Introduction

The previous chapters have dealt with functions invariant to the various geometric image transformations caused by object motion, and have only discussed recognition schemes in passing. This chapter discusses various approaches to using the invariant functions to classify objects. In particular, the schemes are designed to cope with partial occlusion — very often, an object needs to be recognized when only part of it is visible, a task which we humans usually accomplish with ease.

Once invariants have been extracted from an image, two steps are performed in the schemes outlined below: the invariants are used to index a table containing the names and parts of the objects that may be present, and then model views of these objects are 'back-projected' onto the image to see whether they fit. The chapter starts with a discussion of two indexing schemes, one global and the other local; the next section discusses how well-suited different invariants are for use with either of the schemes, and then section 6.4 shows how to perform back-projection using affine and projective transformations — in particular, a technique for affine back-projection based on an identified section of the object's boundary is presented.

The results of some experiments using images of coarsely sampled near-planar objects (a spanner, a hammer, a file and a chuck-key), in which the effects of perspective are noticeable, are presented in section 6.5. Forsyth *et al.* [9] and Coelho *et al.* [109] show that projective invariants are stable when computed for finely sampled planar objects, and Lamdan *et al.* [21, 19] and Costa *et al.* [86] indicate that affine point-based invariants can reliably recognize near-planar objects in finely sampled images, but neither investigates the effects of coarse sampling. Four invariants are investigated: the projective invariant based on two lines and two points; the invariant affine coordinates; the point-based moment invariants and the local moments. In addition to investigating the stability of these invariants, the section discusses their effectiveness in the relevant indexing schemes and shows the effectiveness of affine back-projection based on curve sections. Section 6.6 concludes the chapter with a summary of the main findings.

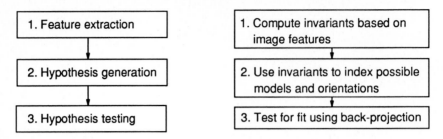

Figure 6.1: The three stages used in object recognition.

On the left, the three recognition stages typically used in computer vision, and on the right their implementation when using invariant features.

6.2   Recognition schemes

This section discusses various approaches to recognizing partially occluded objects given a database of model images, and is in two parts. The first part discusses the trade-off between indexing using invariants and searching through the database for a good fit while the second part looks at two different approaches to using invariant features.

6.2.1   Search versus indexing

A brute-force approach to recognizing unoccluded objects extracts features from the image, such as lines and points, and then sequentially searches through the database of model views, trying to find a projection of the model that could produce the given combination of image features [123, 124, 125]. Clearly, if the database is large and the models are fairly complex, the computation involved in recognizing an object will be large; on the other hand, the amount of memory required will be a minimum. The situation becomes much worse when considering partially occluded objects; indeed, the computational complexity goes from being quadratic in the parameters to being exponential in them [126].

Invariant features appear to offer the only way of avoiding the computational load of the brute-force approach. When recognizing partially occluded objects, invariant features are used to index a look-up table, typically implemented using geometric hashing [19, 20]. The values of the invariants are calculated for each object in the database and used to compute a hash value which indexes a bin in which the identity of the model with those invariants is stored. Recognition proceeds in two stages: first, invariants are computed from the image, combined to form a hash value which is then used to determine which features from which models may be present in the image. In the second stage, the selected models are projected onto the image and a measure of fit is computed. Typically, the model that fits best is said to be present in the image as long as its measure of fit is above a certain threshold, (Grimson & Huttenlocher [127] discuss how to choose a suitable threshold). Figure 6.1 shows the three stages used.

Ideally one would like the invariants to index a single model and orientation; we have already seen that indexing a single model is possible when recognizing unoccluded objects. As we will see below, invariants used to recognize partially occluded objects generally provide less discrimination, resulting in a multitude of potential matches between models and image. The choice of invariants allows a trade-off between the complexity of stage two with the complexity of stage three: increasing the complexity of stage two can increase the invariants' discrimination and hence reduce the number of hypotheses that need testing in stage three.

6.2.2 Local versus global indexing

Given that one is going to extract reference features such as points, lines, tangents etc. from the image boundary, how should one best combine them to form invariants that will allow reliable indexing under partial occlusion? Two methods can be used; the first can cope with more occlusion than the second, but typically requires more storage and computation.

The first method, which I shall refer to as global indexing, essentially uses exhaustive search, but can be speeded up considerably by using look-up tables implemented by hashing [19, 128, 129]. In the simple case, one chooses one type of invariant to use, which requires a minimum of N references features (e.g. $N = 4$ for affine coordinates). Next, a look-up table is constructed in a pre-processing stage as follows: assume we have a model with $M > N$ reference features. We take all permutations of $N - 1$ features (called *basis* features) and, for each one, compute $M - (N - 1)$ invariants based on each of the $M - (N - 1)$ remaining features. Each of these invariants, suitably quantized, is used to index the look-up table in which the model identity and the $N - 1$ basis features are stored. Having done this for all models, we are in a position to recognize instances of a model in an image. To do this, one extracts all the reference features (e.g. points) from the image, and then one randomly chooses a set of $N - 1$ features to use as basis features. If there are K reference features in the image, one computes $K - (N - 1)$ invariants, one for each reference feature not in the basis. Each of these invariants is used to index the look-up table, and the number of times each model appears with the same $N - 1$ reference points is noted; the model plus basis that appears significantly more often than the others is chosen as being present. If no model plus basis is chosen, another set of $N - 1$ image features is tried as a basis; this is repeated until a model is found, or all possible basis features have been tried. The latter, worst case, requires $O(K^N)$ steps, but is in practice rare; one can show that, given a reasonable proportion of a single model's reference features among the image features, recognition time is $O(K)$ [19]. The smaller the number of features, N, required the better, because one is more likely to find $N - 1$ image features that belong to a single model when N is small — if not all of the $N - 1$ basis features belong to a single model, no model will be recognized using that basis. The above method can in fact be made more accurate by using probabilities rather than votes; see the paper by Costa *et al.* [86].

Although quite complex, the above scheme has two nice features: first, it is robust to missing reference features — if, for example, an unoccluded reference feature is not detected, it will not affect recognition accuracy as long as a sufficient number of other reference features have been extracted. Second, it uses all the information available in the image, given the type of invariants that are being used — in Lamdan

Affine invariants			Projective invariants		
Invariants	N	k	Invariants	N	k
Moments	1	-	2 points + tangents & curvatures (Θ)	2	1
Joint moments	2	-	Conic + 2 lines/points	3	2
2 lines, point + tangent	3	1	3 points + tangents	3	1
4 points	4	2	2 lines + 2 points	4	1
			3 lines + point & tangent	4	1
			5 points/lines	5	2

Table 6.1: A summary of image invariants.

The table shows the number of features N required to compute each invariant, and the number k of independent features provided by the N features. The moments and joint moments of curve sections can in principle generate an unlimited number of independent features — see the text.

et al.'s case all model points in an image are used to identify the model.

The second scheme, which I shall refer to as local indexing, is much simpler than the first, but has neither of the two aforementioned advantages. In the pre-processing for this scheme, a look-up table is also constructed; each entry corresponds to a reference feature on the model boundary and contains the value of an invariant based on the $N-1$ neighbouring reference features on the boundary. Hence, if one only uses points as features, and chooses the cross-ratio as the invariant, the invariants would be the cross-ratio of the boundary point and the two points either side of it on the boundary. The value of the invariant is used to index a look-up table as above, only this time the model and the position on the boundary is stored. Recognition proceeds by extracting the boundaries of a possible object in the image, and then computing invariants for each of the K reference features and keeping count of the number of times each model is indexed in the look-up table.

Because the K reference features along the boundary are searched in order, the worst case complexity is only $O(K)$. However, because it requires at least N consecutive features on the boundary to be unoccluded, it will fail to classify objects for levels of occlusion that would not affect the first method. Rothwell *et al.* [20] use this technique, but obtain slightly increased robustness by using lines and conics as features — they are less easily occluded than points.

6.3   Choice of invariants

As we have seen in the preceding chapters, we have quite a large number of invariants that we could extract from the image. In this section we will look at the trade-offs involved in choosing which ones to use. Before considering affine and projective invariants separately, let us consider aspects that apply in both cases.

Table 6.1 summarizes the invariant features that can be used under affine transformations and under projective transformations. They are discussed in more detail below; for the moment, note that one can use points, lines, conics, tangents at points, curvature at points, local signatures and local moments. What kind of information do the various features convey about the image, and hence how do they contribute to discrimination between different objects? First, note that when the number of features N required to compute an invariant is large, the hash table will be very large, since the amount of memory required is $O(M^N)$ for an object with M reference features: if the features are sorted, a model with M reference features generates $N_m = M!/(M-N)!$ table entries, whereas if they are unsorted $N_m/N!$ entries are generated. For $N = 4$, $N_m = M(M-1)(M-2)(M-3)$ and $N! = 24$. Hence, if we assume $M = 20$ on average and we have 30 objects, and each entry in the hash table is 4 bytes long, the table will require approximately 13.6 Mbytes of memory — a large amount. If one has $N = 5$, this grows to 218 Mbytes, an unfeasibly large amount. Although unsorted features use less memory, they will always result in at least $(N-1)!$ times as many hypotheses to be tested.

Now, the larger a hash table, the more likely one is to find more than one match to a given image configuration (a property which is independent of the implementation of the hash table). This arises from the fact that each invariant is assigned a margin of error ϵ to account for imaging errors. If we use k independent invariants I_i, $i = 1,\dots,k$, to form an index (e.g. $k = 2$ for the affine coordinates), the k invariants define a point in a k-dimensional space and the error region will be a hyper-cube of side 2ϵ and volume $(2\epsilon)^k$ centred at this point. In most cases one can consider the invariants to be bounded for a given set of models, which means that all index points must lie within some bounded k-dimensional space of volume $V = L^k$, $L \gg \epsilon$ (L defines the dynamic range of the invariants). If we assume that the N_m entries in the hash table corresponding to a model with M features are uniformly distributed in the volume V, then we see that we are bound to have two different configurations with overlapping error hypercubes when $N_m > V/(2\epsilon)^k = (L/2\epsilon)^k$. Since the N_m entries are not uniformly distributed, one will find some overlapping entries much sooner than this. N_m is $O(M^N)$, so increasing N will increase the likelihood of multiple hypotheses, as we set out to show. Similarly, invariants with small L (small dynamic range) will increase the likelihood of multiple hypotheses. By the same argument we see that increasing the number of features k used to form the index will decrease the likelihood of multiple hypotheses.

Having looked at the factors that contribute to multiple hypotheses, let us now look at the properties of various features one can use to obtain invariants. When considering the utility of these features, the reader may find it helpful to look at figures 6.13–6.19 which contain images of the near-planar objects in figure 6.8 occluding one another, along with some of the extracted features.

Point features: The most robust point features are the bitangent points (see figure 1.5) which, as the name suggests, require the extraction of tangents from the boundary. Other robust point features are sharp extrema in curvature, which essentially require second-order derivatives to detect (see section (5.2.3)). Using invariants based on point features to index the hash table means that one is not using any information about the direction and curvature of the boundary at the reference points, nor is one using any information about the shape of the boundary near the reference points. As a result N is relatively large for point features — $N = 4$ for affine invariants and $N = 5$ for projective ones — and so one is likely to generate quite a few hypotheses to test. Using point features postpones the use of curve information to the back-projection stage, where the image curve is compared with a projection of the model curve.

Line features: Lines can generally be extracted reliably; however, they suffer from similar drawbacks to point features: although they carry a lot of information about the shape of a curve, they carry no information about where on the (infinite) line the linear part of the object is to be found. Nevertheless, picking out the linear portions of a curve seems to be a natural way of describing an object; in particular, note how the two lines are very helpful in recognizing the spanner in figure 6.15.

Tangents and curvatures at points: The tangent at a point conveys some information about the shape of the curve at a point, but it says nothing about the way the curve behaves either side of the point. This information is contained in part by the curvature at the point, as can be seen by considering the radius of curvature. Irrespective of whether one is extracting invariants or not, describing a curve by referring to significant points such as sharp extrema of curvature and bitangent points along with the tangent and curvature at these points seems very natural.

Conics: Just as with image lines, it seems natural to extract circles and ellipses from an object boundary if they are present; unfortunately, except with some machine parts, they are not present in most images.

Curve descriptions: These are functions of the whole curve between two reference points, such as functions of the curve's signature or moment invariants of the shape defined by the curve between the two points. Curve descriptions are unlikely to provide much discrimination unless the curve varies in an interesting manner between the two reference points — preferably in a manner unique to the given object. A perfect example of an uninteresting curve is a straight line joining the two points; in contrast, the curve between two bitangent points on the same tangent is likely to be informative. Using curve descriptions gives one the potential to recognize unoccluded objects with $N = 1$, i.e. using local indexing; however, the interesting curve sections can easily be partially occluded, preventing the object from being recognized. Alternatively, one could increase discimination by using a voting scheme based on the curve between a reference point and K other points along the boundary, for some fixed K. Experiments are described below using this idea with affine invariants.

6.3.1   Invariants for affine transformations

Lamdan *et al.* [21, 19] implemented global indexing with points as features; given four points one can very easily compute two affine invariants, so $N = 4$ and $k = 2$, and they search through basis triplets in the image. (See also references [128, 129].) Huttenlocher [130, 87] also used an indexing scheme based on slightly different affine invariants, but his scheme is not very robust to occlusion. One can improve on Lamdan *et al.*'s method on two fronts by using moment invariants (see figure 6.2) — boundary information is not discarded and $N = 2$, resulting in reduced complexity. As depicted in table 6.3, one can also use two lines, a point and its tangent to give one independent invariant and $N = 3$. In principle one can use the moment invariants of the hatched region R_1 in figure 6.2 (a) defined by two successive boundary points or by a point and the kth successive point, $k = 1, \ldots, K$, to identify models. The advantage of doing

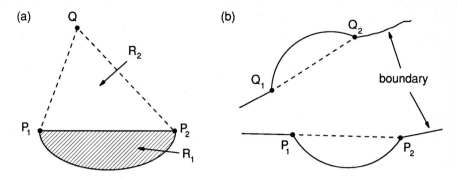

Figure 6.2: Using moment invariants for recognition under occlusion.

(a) The moment invariants of the hatched region, or the joint invariants of the two regions P_1QP_2 and the hatched region can be used as reference features. (b) Q_1, Q_2 and P_1, P_2 are pairs of neighbouring reference points on the boundary. One can use global indexing, with P_1 and P_2 as a basis, and compute moment invariants of the shape $P_1P_2Q_2Q_1$ for all neighbouring pairs of points Q_1 and Q_2, or joint invariants of the two hatched regions; in both cases $N = 2$.

this is that one has the speed of local indexing; the disadvantage is that it is more susceptible to occlusion than schemes using global indexing. Two simple schemes that use global indexing are depicted in figure 6.2. Given a third point Q, one can form joint invariants using the moments of the regions R_1 and R_2. Let us take Q as the origin of the coordinate system, and denote the moments of R_1 by m_{pq} and those of R_2 by μ_{pq}. The simplest joint invariant is the ratio of the two areas, $I_1 = m_{00}/\mu_{00}$. The next simplest are I_2, based on the joint invariant of two linear binary forms, and J_2, based on the invariant of linear and quadratic forms (3.6):

$$I_2 = \frac{m_{10}\mu_{01} - m_{01}\mu_{10}}{(m_{00} + \mu_{00})^3}; \qquad J_2 = \frac{\mu_{10}^2 m_{02} - 2\mu_{10}\mu_{01}m_{11} + \mu_{01}^2 m_{20}}{(m_{00} + \mu_{00})^5}.$$

Further joint invariants can be found by applying the revised fundamental theorem of moment invariants to the joint invariants of higher order binary forms. $N = 2$ in this case, and we can use either the region R_1 or the point Q as the basis.

The drawback with this scheme is that the region R_1 is undefined when the boundary between P_1 and P_2 is a straight line. This can be overcome by replacing Q with a pair of neighbouring points Q_1 and Q_2. $N = 2$ as before, but this time we can compute the moment invariants of the region $P_1P_2Q_2Q_1$. Alternatively, one can generalize the former scheme, exclude linear curve segments and use the joint invariants of the two curve regions in figure 6.2(b), giving $N = 2$ again. Since the curve regions are typically considerably smaller than the whole object, the affine approximation to projection will hold more nearly than when using $P_1P_2Q_2Q_1$.

These schemes require one or two pairs of neighbouring model reference points to be correctly identified on the image, which makes them slightly more susceptible to occlusion than Lamdan *et al.*'s method; however, since $N = 2$, the worst case number of operations drops from $O(K^4)$ to $O(K^2)$, and one is more likely to find a basis of two points that coincide with an object's reference points than one is when choosing four points at random. Instead of using moments in a global scheme, one can also use them in a local scheme by simply computing the moment invariants of the region R_1 between two neighbouring reference points. This can be generalized as discussed above under 'curve descriptions'.

6.3.2   Invariants for perspective transformations

To obtain invariance under all viewing positions, one must use projective invariants. Table 6.3 lists the invariants that one could use, along with their values of N, the number of reference features required to compute them and k, the number of independent invariants provided by the N image features. The most promising invariant appears to be Θ, since $N = 2$. The fact that one requires second-order derivatives does not necessarily make the scheme infeasible, since one needs to compute these to detect the reference points defined by extrema of curvature. The invariant of three points and their tangents and that of a conic and two lines or points have potential, with $N = 3$, as does the invariant of two lines and two points — although $N = 4$, the number of hash-table entries will be lower than when using four points because the number of straight lines detected on most objects will be significantly lower than the number of points.

Just as one can use two pairs of points P_1P_2 and Q_1Q_2 to get an $N = 2$ scheme with moments for the affine case, one can use two pairs of points with van Gool *et al.*'s signatures [131] to generate invariants (see section 5.2.3). As with the moments, $N = 2$ and we are using information of the shape of the entire curve between the pairs of reference points. Again as with the moments, a feature that consists of a pair of points is more likely to be occluded than a single point. Van Gool *et al.*'s signatures can also be used with local indexing, fulfilling the same function as the moment invariants in the previous section.

It remains to be seen which of the above invariants performs best in practice, and whether one can usefully combine information from more than one invariant in order to reduce the number of hypotheses. Some preliminary experimental results are presented in section 6.5.

6.4   Back-projection

Having seen how to generate hypotheses, we will now consider how to project the model to its hypothesized orientation in the image. We will look at the affine case first, followed by the projective one. References [125, 123, 124] provide alternative approaches to back-projection.

6.4.1   Affine back-projection

Three non-collinear reference points on image and model suffice to determine the affine transformation linking the two. Jacobs [132] investigates an approach to finding the best match given three reference points. However, because of imaging and segmentation errors, a better fit is obtained if one uses more than three points and computes the affine transformation that minimizes the squared distance between all the points.

Lamdan *et al.* [19] use this in their scheme; one can either use the equations in their paper, or one can use the following more concise method.

Let the reference points on the model have coordinates (X_i, Y_i), $i = 1, \ldots, N$, and the points to which they are affinedly transformed have coordinates (X_i', Y_i'):

$$\begin{bmatrix} X_i' \\ Y_i' \\ 1 \end{bmatrix} = \begin{bmatrix} a_{11} & a_{12} & a_{13} \\ a_{21} & a_{22} & a_{23} \\ 0 & 0 & 1 \end{bmatrix} \begin{bmatrix} X_i \\ Y_i \\ 1 \end{bmatrix}. \tag{6.1}$$

Further, assume that the N model points (X_i, Y_i) correspond to image points $(\hat{X}_i, \hat{Y}_i)$; our task is to minimize the sum of the squared distances E between the N points $\{(\hat{X}_i, \hat{Y}_i)\}$ and the N points $\{(X_i', Y_i')\}$:

$$E = \sum_{i=1}^{N} (\hat{X}_i - X_i')^2 + (\hat{Y}_i - Y_i')^2. \tag{6.2}$$

If we write (6.1) as $\mathbf{v}_i' = \mathbf{A}\mathbf{v}_i$, and define $\hat{\mathbf{v}}_i^T = [\ \hat{X}_i\ \ \hat{Y}_i\ \ 1\ ]$, then

$$E = \sum_{i=1}^{N} \|\hat{\mathbf{v}}_i - \mathbf{A}\mathbf{v}_i\|^2. \tag{6.3}$$

If we define $\mathbf{V} = [\ \mathbf{v}_1\ \ \mathbf{v}_2\ \ \cdots\ \ \mathbf{v}_N\ ]$ and $\hat{\mathbf{V}}$ similarly, then we can show in a few lines that E is minimized using the pseudo-inverse of $\mathbf{V}$ [121]:

$$E = \|\hat{\mathbf{V}} - \mathbf{A}\mathbf{V}\|^2 \quad \Rightarrow \quad \min_{\mathbf{A}} \{E\} = \mathbf{A}_{\min} = \hat{\mathbf{V}}\mathbf{V}^T(\mathbf{V}\mathbf{V}^T)^{-1}.$$

(Note that $\hat{\mathbf{V}}\mathbf{V}^T$ and $\mathbf{V}\mathbf{V}^T$ are both 3×3 matrices.)

The above considers the case when we have four reference points on image and model; how does one proceed when recognizing an object using local moments of the boundary between two reference points? One could use the moment normalization equations of section 4.3; however, experiments performed by the author indicate that the following technique is simpler and more accurate, because it does not suffer from problems with rotational symmetries. If one uses the midpoint between the two reference points as the origin (this choice is invariant to affine transformations), one can generate two further reference points on the boundary curve between the two end points by selecting them so as to partition the area between the object boundary and the line joining the end points into three equal areas — see figure 6.3. Together with the two original reference points we now have four reference points which suffice for back-projection as discussed above.

6.4.2 Projective back-projection

The above technique can be generalized to deal with projective transformations. Given four reference points on image and model, one can determine the projective transformation linking the two. As in the affine case, a more accurate fit is possible if one uses more points and minimizes a least-squared distance. Below we will first see how one can determine the projective transformation given four points, and then how to

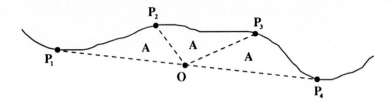

Figure 6.3: Generating four reference points given two points on the boundary.

P_1 and P_4 are the original reference points. The origin O is chosen to be the midpoint between P_1 and P_4, and P_2 and P_3 are points on the image boundary that partition the region enclosed by the image boundary and the line joining P_1 and P_4 into three regions with equal area A.

generalize the method to give a novel quasi least squares fit which has an analytical solution. The same notation as in the affine case is used throughout. An example is given in which the least-squares approach can be seen to be robust.

First, consider the case where we have four reference points. As we saw in chapter 1, the projective transformation linking a point (X', Y') on one plane with a point (X, Y) on another plane is given by

$$X' = \frac{a_{11}X + a_{12}Y + a_{13}}{a_{31}X + a_{32}Y + 1}, \qquad Y' = \frac{a_{21}X + a_{22}Y + a_{23}}{a_{31}X + a_{32}Y + 1}. \tag{6.4}$$

One can obtain the following vector equation by multiplying both sides of (6.4) by the denominator $a_{31}X_i + a_{32}Y_i + 1$:

$$\hat{\mathbf{P}}_i\mathbf{a} = \begin{bmatrix} \hat{X}_i \\ \hat{Y}_i \end{bmatrix}, \qquad i = 1, 2, 3, 4, \tag{6.5}$$

with $\quad \hat{\mathbf{P}}_i = \begin{bmatrix} X_i & Y_i & 1 & -X_i\hat{X}_i & -Y_i\hat{X}_i & 0 & 0 & 0 \\ 0 & 0 & 0 & -X_i\hat{Y}_i & -Y_i\hat{Y}_i & X_i & Y_i & 1 \end{bmatrix}$

and $\quad \mathbf{a}^T = \begin{bmatrix} a_{11} & a_{12} & a_{13} & a_{31} & a_{32} & a_{21} & a_{22} & a_{23} \end{bmatrix}$.

Equation (6.5) gives us eight linear equations in eight unknowns, which can be written as

$$\hat{\mathbf{P}}\mathbf{a} = \mathbf{d} \qquad \Rightarrow \qquad \mathbf{a} = \hat{\mathbf{P}}^{-1}\mathbf{d},$$

where $\hat{\mathbf{P}}^T = [\hat{\mathbf{P}}_1^T\ \hat{\mathbf{P}}_2^T\ \hat{\mathbf{P}}_3^T\ \hat{\mathbf{P}}_4^T]$ and $\mathbf{d}^T = [\hat{X}_1\ \hat{Y}_1\ \hat{X}_2\ \hat{Y}_2\ \hat{X}_3\ \hat{Y}_3\ \hat{X}_4\ \hat{Y}_4]$.

Now to generalize this to least squares back projection. Ideally we would like to minimize the sum of the squared errors E as in equation (6.2), only this time we have

$$\begin{bmatrix} X_i' \\ Y_i' \end{bmatrix} = \mathbf{P}_i\mathbf{a}, \qquad i = 1, \ldots, N, \tag{6.6}$$

with $\quad \mathbf{P}_i = \begin{bmatrix} X_i & Y_i & 1 & -X_iX_i' & -Y_iX_i' & 0 & 0 & 0 \\ 0 & 0 & 0 & -X_iY_i' & -Y_iY_i' & X_i & Y_i & 1 \end{bmatrix}. \tag{6.7}$

Minimizing E is now nonlinear; however, we can get a linear approximation by replacing X_i' and Y_i' in $\mathbf{P}_i$ (6.7) with the image values $\hat{X}_i$ and $\hat{Y}_i$. If we call the resulting

$\mathbf{P}_i$ matrix $\tilde{\mathbf{P}}_i$ and the resulting coordinates $(\tilde{X}_i, \tilde{Y}_i)$, (6.6) becomes

$$\begin{bmatrix} \tilde{X}_i \\ \tilde{Y}_i \end{bmatrix} = \tilde{\mathbf{P}}_i \mathbf{a} \qquad i = 1, \ldots, N.$$

We can now proceed to minimize the squared distance between these new points $(\tilde{X}_i, \tilde{Y}_i)$ and the image points $(\hat{X}_i, \hat{Y}_i)$:

$$E = \sum_{i=1}^{N} \left(\hat{X}_i - \tilde{X}_i \right)^2 + \left(\hat{Y}_i - \tilde{Y}_i \right)^2 \tag{6.8}$$

$$= \sum_{i=1}^{N} \| \hat{\mathbf{d}}_i - \tilde{\mathbf{P}}_i \mathbf{a} \|^2 \tag{6.9}$$

$$= \| \hat{\mathbf{d}} - \tilde{\mathbf{P}} \mathbf{a} \|^2, \tag{6.10}$$

where $\hat{\mathbf{d}}_i^T = [\hat{X}_i, \hat{Y}_i]$, $\hat{\mathbf{d}}^T = [\mathbf{d}_1^T \, \mathbf{d}_2^T \cdots \mathbf{d}_N^T]$ and $\tilde{\mathbf{P}}^T = [\tilde{\mathbf{P}}_1^T \, \tilde{\mathbf{P}}_2^T \cdots \tilde{\mathbf{P}}_N^T]$. As before, E is minimized by using the pseudo-inverse to choose $\mathbf{a}$:

$$\mathbf{a} = \tilde{\mathbf{P}}^\# \hat{\mathbf{d}}, \qquad \tilde{\mathbf{P}}^\# = (\tilde{\mathbf{P}}^T \tilde{\mathbf{P}})^{-1} \tilde{\mathbf{P}}.$$

Although this only approximately minimizes the least-squared distance between the projected model points and the image points, it is exact if the N model points can be exactly projected onto the N image points. Figure 6.4 shows the improvement in accuracy that the least-squares approach achieves over the exact one using four points.

If one has extracted features other than points from the image, such as lines and tangents, one can apply the above by simply finding points where lines and tangents intersect and, if need be, creating further lines by joining known points. Doing this for three points and their tangents gives us a total of nine points (note that only nine of the eighteen coordinates are independent, since three points and their tangents only have nine degrees of freedom). The situation is more complicated if one of the features used is a conic; Forsyth *et al.* [9] show how to perform back-projection in this case.

6.5 Experimental results

The experiments discussed below are designed to investigate how robust affine and projective invariants are to the distortions caused by slight non-planarity and coarse sampling. The objects used are a spanner, a hammer, a file and a chuck-key, and can be seen in figure 6.8 and in figures 6.9–6.12. The four invariants investigated are the projective invariant of two lines and two points, the invariant affine coordinates, the point-based moment invariants and the local moments; the section on the latter also investigates the accuracy of the affine back-projection method based on two points on the boundary. Before discussing the results, mention should be made of the type of reference points used. Three types of bitangent points can be extracted from an image: those based on tangents that touch the object on the outside in both cases ('external-external' points, see figure 6.5), those whose tangent touches once on the inside and once on the outside (figure 6.6) and those whose tangent touches internally both times

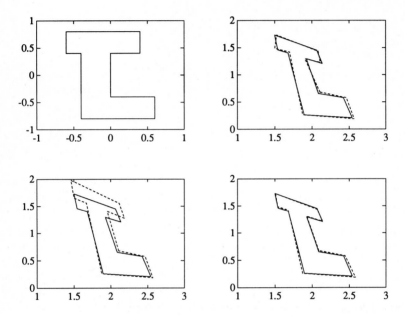

Figure 6.4: Projective back-projection.

Top left: *Image stored in database.* Top right: *The shape with the solid outline is a projection of the image top left. The shape with the dashed outline is created by randomly perturbing the vertices of the shape with the solid outline.* Bottom row: *The shape with the solid outline is the same as in the figure top left.* Bottom left: *The shape with the dashed outline is the exact projection of the original image assuming the bottom four vertices must project onto the corresponding bottom four vertices of the distorted shape top right.* Bottom right: *The shape with the dashed outline is the projection of the original image using pseudo-least-squares on all corresponding ten vertices of the distorted shape top right. Pseudo-least-squares is clearly superior.*

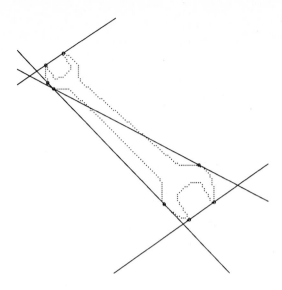

Figure 6.5: Examples of 'external-external' bitangent points.

(figure 6.7). In the following experiments only 'external-external' bitangent points have been extracted as reference points; although one generally obtains more reference points per model using the other bitangent points too, this will be counteracted by obtaining far more reference points in a cluttered image — the fewer the number of reference points in the image, the swifter recognition should be. In all cases the reference points were extracted manually; they can be extracted automatically, as witnessed by C.A. Rothwell's bitangent extracting program at Oxford University.

In addition to extracting 'external-external' bitangent points, significant pairs of lines were extracted from each object, using a least-squares fit, to allow the use of the projective invariant of two lines and two points. The extracted lines are shown in the figures as solid lines; the extracted reference points are shown as crosses and circles, where the circles are simply those points that are used as the origin when computing the local moment invariants — this is discussed further below. Figures 6.9–6.12 show the original (model) views of each object along with the extracted features, and figures 6.13–6.19 show views of the objects partially occluding one another, along with the extracted reference points and lines that correspond to those in the model views. (N.B. The points on the model views were selected to correspond with unoccluded bitangent points in the occluded images.)

Three of the invariants tested would need to be combined with global indexing: the projective invariant of two lines and two points, the affine coordinates and the point-based moment invariants. The fourth set of invariants tested, the local moment invariants, would be combined with local indexing. When comparing the performance of each set of invariants, one must be aware of the trade-offs involved. First, as long as one vote is cast for the correct basis, the correct object should eventually be identified because of the back-projection step. However, one would like to minimize the number

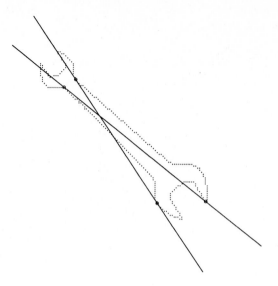

Figure 6.6: Examples of 'external-internal' bitangent points.

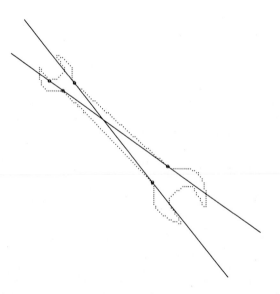

Figure 6.7: Examples of 'internal-internal' bitangent points.

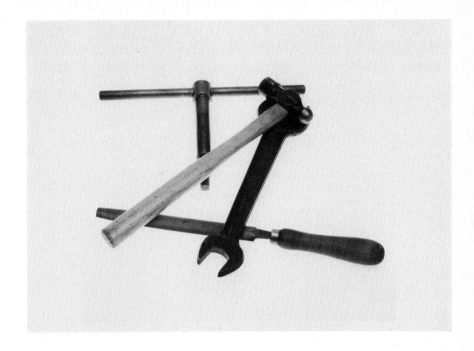

Figure 6.8: The four tools used in the experiments.

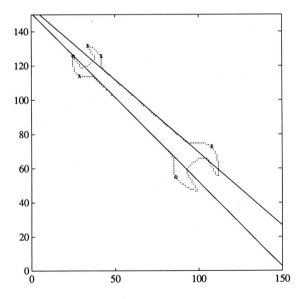

Figure 6.9: Original view of spanner with extracted features.

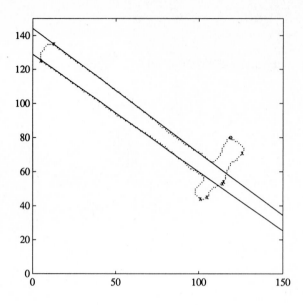

Figure 6.10: Original view of hammer with extracted features.

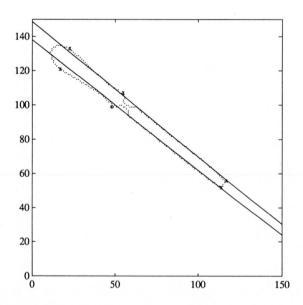

Figure 6.11: Original view of file with extracted features.

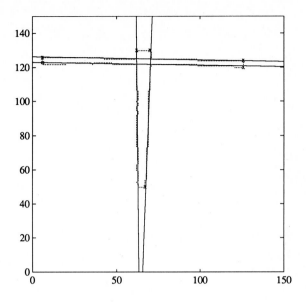

Figure 6.12: Original view of chuck-key with extracted features.

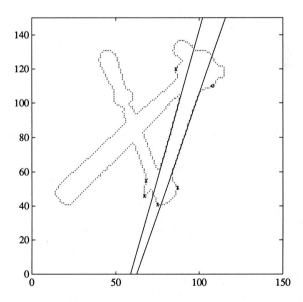

Figure 6.13: Occluded view A with extracted features of the spanner.

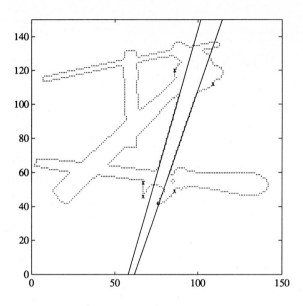

Figure 6.14: Occluded view B with extracted features of the spanner.

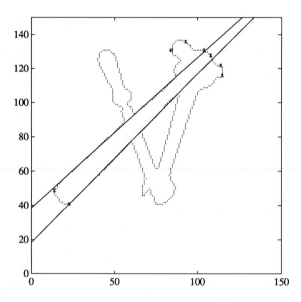

Figure 6.15: Occluded view A with extracted features of the hammer.

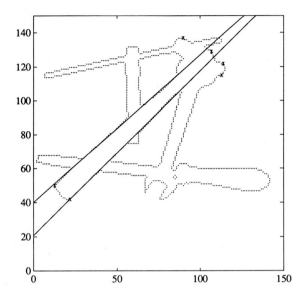

Figure 6.16: Occluded view B with extracted features of the hammer.

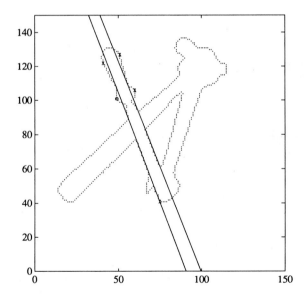

Figure 6.17: Occluded view A with extracted features of the file.

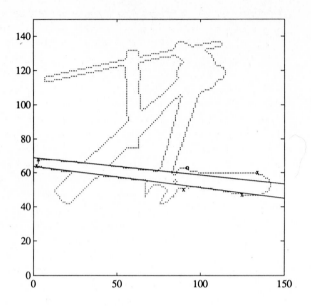

Figure 6.18: Occluded view B with extracted features of the file.

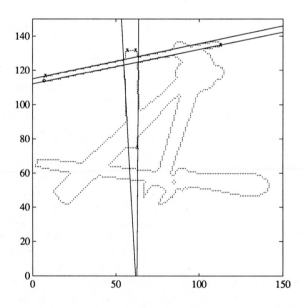

Figure 6.19: Occluded view B with extracted features of the chuck-key.

of false hypotheses, i.e. the number of back-projections required before the correct match is found. A rough guide to the number of 'false votes' is given by the number of entries per bin in the table, so one would like to keep this number as small as possible. A rough estimate of this number can be obtained as follows: if an object's model view contributes N_e entries to the database, and we assume that they are uniformly distributed over the range of each of the k invariants, each of which is partitioned into b bins, then we obtain a total of b^k bins and we expect about N_e/b^k entries per bin. In practice the distribution will not be uniform, so some bins will have more entries than this, but the above analysis provides a good rule of thumb. In the tables, the error margin ϵ was selected to give a compromise between the number of entries per bin and the number of votes cast. Note that in the experiments described below, the invariants are *not* quantized into bins; instead, they must be within ϵ of the value in the original (model) view for a vote to be cast.

Another consideration that is important when wanting to recognize a large number of objects is the size of the database required. These issues are discussed below together with the experimental results for each of the four sets of invariants.

6.5.1   Invariant of two lines and two points

Before looking at the results of table 6.5.1 for the invariant of two lines and two points, let us look at the likely structure of the hash table. The spanner shown in figure 6.9 shows only eight of the external-external bitangent points; four more can be extracted, giving a total of twelve. The invariant depends on the labelling of the lines and points, so the spanner will contribute a total of $2 \times 1 \times 12 \times 11 = 264$ entries to the hash table. Because the invariant can take on very large, but not very stable, values, values above 2 and below -2 were placed in the bins at the extreme of the scale. Choosing the error margin $\epsilon = 0.05$ thus gives 40 bins, so we expect around 6.6 entries per bin from the spanner alone. If every model contributes as many entries, we see that one is likely to perform many back-projections, unless an object obtains quite a few votes (e.g. three or more).

Table 6.5.1 shows the values of the invariant for the different views of the four objects, along with the number of votes for each partially occluded view. A vote is cast every time an invariant is within ϵ of its value in the original view 'O'. With the exception of the hammer and one set of invariants for the chuck-key, we see that only a small number of the reference points result in votes for the object; nevertheless, in all cases at least one vote is cast, indicating that each object should be correctly identified, albeit after many back-projections.

6.5.2   Invariant affine coordinates

The error margin $\epsilon = 0.3$ over the range ± 3 results in 10 bins per invariant; since there are two invariants, $k = 2$, giving a total of $10^2 = 100$ bins. The affine coordinates depend on the labelling of the points, so the 12 reference points of the spanner will contribute $12 \times 11 \times 10 \times 9 = 11{,}880$ entries to the hash table, giving 119 entries per bin. This is a large number, and will result in a large number of back-projections unless many votes were cast for a single object; however, as table 6.5.2 shows, fewer votes were cast using affine coordinates than using the invariant based on two lines

Object		Points							Votes
Spanner	O	0.023	1.220	0.187	0.236	1.005			
	A	-0.194	1.099	0.159	0.204	0.901			2
	B	-0.099	1.044	0.169	0.208	0.877			2
Hammer	O	0.024	44.064	0.378	0.950	-0.479	-0.231	0.335	
	A	-0.003	-5.608	0.395	0.968	6.243	-0.358	0.308	4
	B	0.067	-2.937	0.430	1.037	13.109	-0.281	0.342	3
File	O	1.337	23.300	-0.447	28.956				
	A	0.974	15.457	-0.128	14.005				2
	B	1.569	1.576	0.961	2.329				1
C.-Key 1	O	-2.033	0.073	0.037	-1.046	-2.016			
	B	-0.457	0.018	0.085	-0.326	-0.454			1
C.-Key 2	O	0.007	-75.765	17.586	-0.176	-0.019			
	B	0.007	-273.194	3.219	0.089	-0.057			4

Table 6.2: The invariants based on two lines and two points.

To recognize the objects using this invariant, global indexing would be needed, and one would need a sufficient number of votes for an object to be able to recognize it reasonably quickly. If one chooses ϵ, the error margin, as 0.05, one obtains the number of votes given in the right column. The values are given to 3 decimal places; the values of the invariants use the bases described in the text, and the remaining points shown in the figures. 'O' is the original view, 'A' the view of three objects, 'B' the view of four objects. The two results for the chuck-key refer to using the two pairs of parallel lines in the basis.

Obj.		Points										
		ξ	η	ξ	η	ξ	η	ξ	η	ξ	η	
Spnr	O	-2.73	2.70	-0.80	1.67	0.92	-0.03					
	A	-2.69	2.65	-1.22	2.02	1.03	-0.14					2
	B	-2.53	2.50	-0.96	1.82	1.11	-0.20					3
Hamr	O	0.98	2.83	0.91	-0.35	1.02	1.60	1.01	1.71	1.00	2.61	
	A	1.00	2.47	0.92	-0.27	1.02	0.96	1.03	1.37	1.03	2.07	1
	B	0.99	2.64	-	-	-	-	1.05	1.50	1.04	2.32	3
File	O	3.13	-2.12	1.07	-0.87	3.15	-2.52					
	A	3.92	-3.18	1.02	-1.10	-	-					1
	B	3.63	-3.10	1.19	-1.15	3.68	-3.45					1
CKey	O	1.14	-0.14	1.95	-1.13	-0.04	1.00					
	B	1.07	-0.10	1.87	-1.11	-0.07	1.03					3

Table 6.3: The invariant affine coordinates ξ and η.

See previous table for an explanation. The entries show ξ and η given to 2 decimal places, and ϵ for the voting is 0.3. Such an error margin produces only $10^2 = 100$ bins over the range seen in the table, and yet only the spanner and chuck-key, along with one view of the hammer, produce a reasonable number of votes (shown in the right hand column).

and two points. As before, at least one vote was cast in all cases, so each object should eventually be correctly identified.

6.5.3 Point-based moment invariants

The error margin $\epsilon = 0.004$ for invariant Q and $\epsilon = 0.0008$ for I gives 5 bins each, and hence a total of $5^2 = 25$ bins. Since the value of the invariant is independent of the labelling of the points, the spanner will contribute $4! = 24$ times fewer table entries than when using affine coordinates i.e. 495, which corresponds to about 20 entries per bin — still a large number. Because the invariants do not depend on the labelling, one needs to try all the permutations when back-projecting; if two or more votes are cast for a three-point basis, one needs to try the $3! = 6$ possible labellings of the three basis points, but if only one vote was cast, one does not know which three of the four points to treat as the basis so one must try $4! = 24$ possible labellings. With the above number of bins, this means that one is likely to need at least as many back-projections as when using affine coordinates, but at least the database is smaller by a factor of 24.

The results in table 6.5.3 show that the point-based moment invariants are indeed less stable than the affine coordinates, with one view of the hammer and one view of the file obtaining no votes at all, and in most cases the number of votes being lower than when using the affine coordinates.

6.5.4   Local moment invariants

Indexing

Using moment invariants of sections of the boundary between reference points differs from the previous cases in two respects: first, one can use local indexing and second, one is not constrained to use only one or two invariants. The local indexing scheme envisaged here works as follows: for each reference point p_i in the image, compute the moment invariants for the shape defined by the boundary between point p_i and point p_{i+m} and the straight line joining p_{i+m} and p_i, for $1 \le m \le M$ (Jiang & Bunke's fast moment computation method is ideally suited to this task [51]). The experiments assume $M \ge 5$; the larger M, the more global information is incorporated. Each invariant is used to index a table (c.f. global indexing with $N = 2$); however, because the invariants based on a particular starting point are in order of their position along the boundary, votes are only counted for consistent orderings along the image boundary.

Other objects in the image can produce spurious bitangent points, so when classifying an image one should compute the invariants for $M' > M$ points along the boundary from a given starting point, and then count votes. Values that should work well in practice are $M = 6$ and $M' = 10$, but this is only a conjecture.

If we take $M = 6$, we see that the spanner contributes $12 \times 6 = 72$ entries to the table. As mentioned earlier, we are not limited to one or two invariants when using moments. For instance, we can choose k, the number of invariants, to be quite large (e.g. $k = 12$), and then use only two bins for each invariant — this still gives us $2^k = 4096$ bins in total, i.e. the spanner would contribute approximately 0.018 entries per bin. Even with a large number of models in the database, a single vote should result in a small number of back-projections.

The above analysis using bins is only an approximation, and it breaks down slightly when using only two bins per invariant. Rather than quantizing each invariant into two bins, the error margin ϵ has been used to classify objects according to whether each invariant was within a distance ϵ of the model value. However, if the model value is near the upper or lower limit of the chosen range, the chances of a uniformly distributed random sample having a value within ϵ of it will be between $1/2$ and $1/4$. Assume the lower limit is 0 and the upper limit 2, then $\epsilon = 0.5$ will nominally give two bins. However, the probability that a random sample is within ϵ of the model's value will only be $1/2$ when the model's value v satisfies $1/4 \le v \le 3/4$; when $v = 0$ or $v = 2$ the probability will drop to $1/4$. Hence the overall probability of a random sample being within ϵ of v is $\frac{1}{2}[\frac{1}{2} + \frac{1}{2}(\frac{1}{2} + \frac{1}{4})] = \frac{7}{16}$, assuming v and the random sample are uniformly distributed. This corresponds to $16/7$ or approximately 2.29 bins. Most of the invariants are independent, and those that are not (the irreducible ones) will behave as if they are because they are highly nonlinear functions of the others, so with $k = 12$ we get the probability of a chance match as being $(7/16)^{12} = 4.9 \times 10^{-5}$, corresponding to 20,336 nominal bins. This can be implemented using a sorted table of entries rather than a hash table.

In practice, the invariants are unlikely to be uniformly distributed between their maxima and minima; nevertheless, being able to use $k = 12$ will result in considerably fewer back-projections as long as at least one vote is cast. Table 6.5.4 shows the local moment invariants for the spanner, table 6.5.4 shows the twelve chosen invariants for the hammer and table 6.5.4 shows them for the file, along with the values of ϵ

corresponding to two bins. Two points need to be made: first, the number of votes cast is generally high, implying that only one back-projection is likely, and second, the invariants of the spanner vary considerably less than ϵ, indicating that the reference points are stable. Because of the degree of occlusion, only one vote is achievable in both occluded images of the spanner. The invariants behaved rather unstably for the chuck-key, but this is because Bamieh & de Figueiredo's [76, 51] method for computing the moments was used, which gives the outside of an object a different sign from the inside, resulting in the areas m_{00} of the shapes based on the chuck-key being close to zero. This can easily be avoided by checking for shapes with some parts inside the object and some outside, labelling them correspondingly and then treating inside and outside parts as both being inside.

These results indicate that the moment invariants have the potential to provide a robust means of obtaining fast recognition under occlusion; below we will see that back-projection also works well. It remains to be seen how well the local moments will perform in an actual recognition system. Furthermore, it would be interesting to compare the performance of the moment invariants with that of affine and projective signatures based on the curve between two points, along with the projective invariant Θ of two points, their tangents and their curvatures.

Backprojection

The affine back-projection method based on two reference points on a curve segment (see section 6.4) was tested using the hammer, the spanner and the file seen in figures 6.13 to 6.19. The results of the back-projection are shown in figures 6.20 to 6.24; it is generally successful, although the non-planarity of the spanner is clearly causing problems in figure 6.22. This is not surprising since the theory only applies to planar objects; to use it on such near-planar objects one should perhaps use Lowe's 3D back-projection [124] and use the affine back-projection to give a first estimate of the required 3D model transformation using the equations in appendix C.

6.6 Conclusions

In this chapter we have looked at the theory of global and local indexing, the pros and cons of various invariants in conjunction with both types of indexing, back-projection techniques and the results of experiments testing the usefulness of four types of invariants.

Global indexing has the advantage over local indexing that it allows recognition in circumstances when local indexing fails; however, this is at the expense of requiring a very much larger database (hash-table) — the more so the larger the number N of features required to compute an invariant, since a model with K reference features contributes $O(K^N)$ entries to the hash-table. For this reason the invariant based on two points, their tangents and their curvatures, with $N = 2$, looks particularly promising; however, it remains to be seen whether curvatures can be extracted sufficiently accurately from discrete images.

The stability of an invariant is obviously important for it to be useful, but so is its dynamic range; the two together determine the effective number b of 'bins' the invariant can partitioned into. If a set of features produce k invariants, we will

Object		Points									
		Q	I	Q	I	Q	I	Q	I	Q	I	
Spanner	O	003	057	224	106	140	114					
	A	004	056	215	100	081	102					2
	B	008	052	298	113	076	100					1
Hammer	O	001	061	016	053	118	021	091	026	004	056	
	A	008	052	007	066	447	000	208	011	035	040	0
	B	003	057	-	-	-	-	154	017	016	047	2
File	O	183	120	279	007	062	090					
	A	086	101	407	001	-	-					0
	B	047	089	312	004	017	080					1
C.-Key	O	150	115	252	105	169	117					
	B	149	114	232	095	147	114					2

Table 6.4: The moment invariants of points.

The entries correspond to the 3 digits after '0.2' for the affine invariant based on Q and the 3 digits after '-0.0' for the invariant based on I. The votes were computed using $\epsilon = 0.004$ for Q and $\epsilon = 0.0008$ for I; each value gives a total of $5^2 = 25$ bins over the range shown in the table.

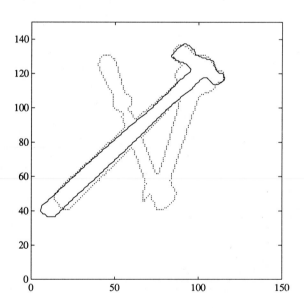

Figure 6.20: Back-projecting the hammer.

Inv.	10^x	Long segment		Short segment	
View:		O	A	O	A
ψ_1	-3	56	56	49	55
ψ_2	-5	79	78	46	69
ψ_3	-5	53	55	53	56
ψ_4	-5	58	56	32	50
ψ_5	-3	21	20	15	19
ψ_6	-5	42	42	28	37
ψ_7	-3	18	18	14	17
ψ_8	-5	31	32	21	28
ψ_9	-5	26	26	17	23
ψ_{10}	-6	21	21	12	18
J_1	-3	49	49	42	48
J_2	-5	-4.2	-11	-44	-46
Γ_1	-2	73	73	69	72
Γ_2	-1	60	57	45	54
Γ_3	-3	51	54	81	61
Γ_4	0	39	37	28	34
Γ_5	-2	53	54	61	54
Γ_6	0	35	34	26	31
Γ_7	-2	39	40	45	41
Γ_8^{-1}	-2	51	52	64	57

Table 6.5: Moment invariants of segments of the spanner.

The value in the second column indicates that the top left entry in the table has the value 56×10^{-3}; likewise for the other entries. The invariants are the moment invariants computed using the long segment on the original image and in the image with three objects, and the short segment on the original and in the image with four objects. Each segment obtained a vote (see table 8.7 for the values of ϵ for each invariant.

Inv.	10^x			End points			
ψ_1	-3	O	73	93	101	75	69
		A	73	68	97	65	58
ψ_2	-5	O	210	580	782	250	183
		A	207	172	634	147	92
ψ_3	-5	O	40	-23	-44	34	43
		A	41	34	-12	51	53
ψ_4	-5	O	163	446	599	190	138
		A	162	132	485	109	67
ψ_5	-3	O	39	81	99	45	36
		A	38	34	81	30	23
ψ_6	-5	O	71	143	158	91	77
		A	72	59	118	58	43
ψ_7	-3	O	32	58	68	35	30
		A	32	27	59	25	20
ψ_8	-5	O	53	105	117	67	57
		A	55	44	86.	43	32
ψ_9	-5	O	48	106	121	60	50
		A	49	39	84	36	26
ψ_{10}	-6	O	47	120	147	62	49
		A	48	37	103	34	23
J_1	-3	O	65	82	89	66	61
		A	65	60	85	57	50
J_2	-5	O	-324	-670	-653	244	73
		A	-210	-183	-215	261	-62

Table 6.6: Moment invariants of parts of the hammer.

The value in the second column indicates that the top left entry in the table has the value 73×10^{-3}; likewise for the other entries. The invariants are computed for the original view 'O' and occluded view 'A'. Using the values for ϵ given in the table 8.7 for all 12 invariants results in 3 votes being cast in favour of the hammer; the two points that do not obtain votes are points 2 and 3 — point 2 in 'A' is clearly a different part of the hammer than in 'O', so one would not expect it to obtain a vote. Point 3 is only rejected by one of the twelve invariants, namely J_2.

Inv.	10^x		End points			ϵ	Inv.	10^x		End points			ϵ
ψ_1	-3	O	71	111	59	20	ψ_7	-3	O	31	71	19	15
		A	94	108	66				A	58	68	25	
		B	75	98	60				B	36	55	20	
ψ_2	-5	O	220	697	65	200	ψ_8	-5	O	49	184	35	40
		A	593	644	104				A	104	172	53	
		B	271	464	73				B	59	130	38	
ψ_3	-5	O	26	48	89	40	ψ_9	-5	O	44	170	29	40
		A	-24	52	91				A	103	158	45	
		B	14	60	88				B	56	117	32	
ψ_4	-5	O	166	588	44	150	ψ_{10}	-6	O	43	233	23	50
		A	458	540	75				A	120	214	40	
		B	208	385	50				B	55	146	26	
ψ_5	-3	O	41	82	19	20	J_1	-3	O	62	104	52	15
		A	81	78	26				A	83	101	58	
		B	47	63	20				B	66	91	53	
ψ_6	-5	O	67	225	44	50	J_2	-5	O	-200	471	184	300
		A	140	212	68				A	170	403	-86	
		B	80	161	49				B	-158	408	111	

Table 6.7: Moment invariants of parts of the file.

The value in the second column indicates that the top left entry in the table has the value 71×10^{-3}; likewise for the other entries. The invariants are computed for the original view 'O' and occluded views 'A' and 'B'. Using the values for ϵ given in the next table for all 12 invariants results in 2 votes being cast in favour of the file for each view; the point in 'A' that does not obtain a vote is point 1, and the point in 'B' that does not obtain a vote is point 2; in both cases these are reasonable results. ϵ represents the error margin used in the voting.

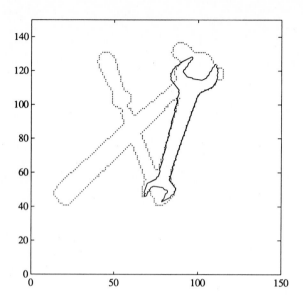

Figure 6.21: Back-projecting the spanner, # 1.

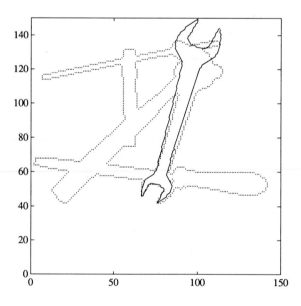

Figure 6.22: Back-projecting the spanner, # 2.

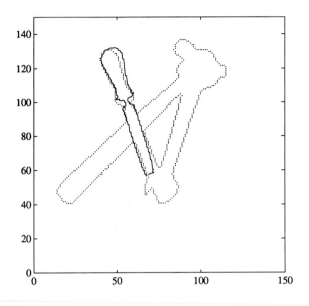

Figure 6.23: Back-projecting the file, # 1.

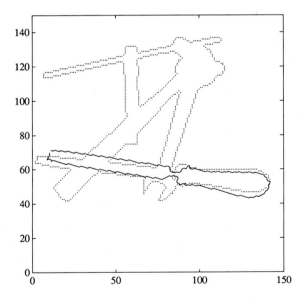

Figure 6.24: Back-projecting the file, # 2.

effectively be quantizing the invariant information from the features into $B = b^k$ bins. Hence, if the hash-table has a total of N_h entries, one will have an average of N_h/B entries per bin. This number will affect the number of false votes, and in turn the number of back-projections necessary to identify an object, and hence the time required for recognition. Thus a combination of invariants and indexing scheme that result in small N_h and/or large B is desirable.

The only type of invariants that allow a large value of k are the signature-type features, such as moment invariants of local boundary segments. The experiments compared their use in a local indexing scheme with that of the projective invariant of two points and two lines, the invariant affine coordinates and the point-based moment invariants using global indexing. Of the latter three, the invariant of two points and two lines performed best; although the number of votes were similar, the invariant of two points and two lines should require far fewer back-projections. All three were outperformed by the use of local moment invariants — more votes were cast, the look-up table is the smallest and the number of back-projections required is likely to be extremely small. The only potential drawback is that local indexing can fail when global indexing using points and/or lines will not.

The results indicate that the theory of invariants of planar objects can be applied to near-planar objects under partial occlusion, and that one can use affine invariants even when the effects of perspective are noticeable. In particular, the local indexing scheme using moment invariants appears to be a very attractive one since it should be quick and only requires a modest amount of storage, and the corresponding back-projection method works well. Of course, the results presented above are only pointers to actual performance; the invariants need to be tested in a fully automatic image recognition package. Nevertheless, the preliminary results described above are essential to narrow down the choice of invariants to use, they are important in establishing that it is worth building a recognition package, and that such a package will be able to recognize partially occluded near-planar objects in coarsely sampled images. In particular, any such package should not just limit itself to projective invariants, since the affine moment invariants perform very well.

Chapter 7

Summary and Conclusions

Throughout we have looked at the use of invariant features to recognize planar objects without having to search through all models in the database, and the main conclusion appears to be that their use is indeed beneficial.

Chapter 2 deals with special cases when the position of the objects to be recognized is constrained, so their images are only likely to undergo one or more of translations, rotations and changes of scale. Image moments or image correlations can be used to obtain invariant features; in the case of moments, one can use invariant features or use normalization to obtain a standard view. Orthogonal moment-based invariants were discussed, and a new orthogonal basis, the Real Weighted Fourier basis, was introduced and shown to outperform the established pseudo-Zernike basis; furthermore, the loss of orthogonality caused by using digital images was analysed and shown to be less severe than expected. In addition to invariance to the above geometrical transformations, invariance to changes in contrast for non-binary images was also discussed.

In most situations, one cannot control the exact location of the objects to be recognized, so one needs features invariant to affine or, even better, to projective transformations. The theory of algebraic invariants lies at the heart of features invariant to both transformations, via image moments for the affine case and via image points, lines and curves in the projective case. Chapter 3 presents a thorough introduction to the theory, and lists a number of invariants, some for the first time; a tutorial on how to generate algebraic invariants is contained in appendix B.

If one assumes that the effects of perspective are negligible (weak perspective), invariance to affine transformations suffices. This is the subject of chapter 4, where a large number of techniques are presented. Image moments can be used to obtain invariant features in two ways, either by using algebraic invariants and the revised fundamental theorem of moment invariants, which is stated and proved, or by normalization, which is presented in the framework of chapter 2. In theory moment invariants allow one to recognize rotational and reflectional symmetries in an object; this was tested, and shown to work well for rotational symmetries but not at all for reflectional symmetry — the only reliable method of discriminating two shapes that are reflections of one another appears to be by normalizing the image and then performing for example template matching. The ability of moments to detect rotational symmetries causes problems for moment-based normalization: high order moments are required in order to normalize objects with rotational symmetries. Moments can also be used to obtain features additionally invariant to changes in contrast, which we saw can be useful even when classifying binary images.

Instead of using moments to obtain invariance to affine transformations, one can use a number of other features, most of which rely on the object's boundary. Examples are Fourier descriptors, point and line based methods, differential invariants, signatures and correlations. Experiments were performed comparing the performance of moment invariants with two types of Fourier descriptors and, separately, with correlation invariants over a wide range of scales, and the moment invariants were found to perform better — they are much more robust to the effects of coarse sampling. The experiments also indicate that the Volterra Connectionist Model (VCM) classifier can give extremely good performance, outperforming the nearest-neighbour algorithm while requiring far less computation. The stability of two point-based affine invariants, the affine coordinates and the moment invariants, was investigated, with the conclusion that the affine coordinates are more stable, but that they incorporate prior knowledge about the labelling of the points. Finally, moment invariants were computed for two coarsely sampled views of a spanner, a hammer, a file and a chuck-key, in which the effects of perspective were noticeable, and found to be reasonably stable.

When the assumption of weak perspective breaks down, we must resort to features invariant to perspective, the subject of chapter 5. If Lenz's analysis [34] were true, we would be in possession of a function of four points that is invariant to the six-parameter perspective transformation; however, since it is not, we must resort to features invariant to the eight-parameter projective transformation. In addition to the projective invariants based on algebraic invariants of polynomial curves presented in chapter 3, one can obtain invariance using differential properties, resulting in invariant signatures and the invariant of two points, their tangents and their curvatures.

In order to use algebraic invariants of polynomial plane curves, one needs to be able to extract the coefficients of a polynomial image curve reliably. A faster method of doing so using algebraic distance was presented for linear transformations, and shown to work well for image conics (i.e. circles and ellipses); however, the invariants of the quartic ternary form were found to be unstable under the effects of sampling, and hence of no practical use [121]. An alternative fitting technique which is projectively invariant was also discussed briefly [111].

Many of the projective invariants rely on reference points extracted from the image; chapter 1 discusses various techniques to obtain such points from an object's boundary, and concludes that bitangent points are the most robust. The stability of the invariant of five points is compared with that of the permutation invariant version in chapter 5, and is found to be much more stable.

Chapter 6 addresses the issue of how to use invariant features to classify planar or near-planar objects in realistic circumstances i.e. under partial occlusion. Global indexing was shown to provide better performance under heavy occlusion, but at the expense of a much larger database. The suitability of various invariants for use with either global or local indexing was discussed, with the conclusion that the invariant based on two lines and two points and particularly the invariant of two points, their tangents and their curvatures are well suited for use with global indexing, and that signature schemes or local moments have an advantage over other invariants for local indexing by allowing a large number of invariant values to be computed from a given curve segment. Furthermore, it was shown how the latter advantages of moment invariants can be combined with global indexing by using joint invariants.

The indexing schemes provide hypotheses about which models may be in the image; these hypotheses are tested by back-projecting the model view onto the image and testing for goodness of fit. Lamdan *et al.*'s [19] point-based affine back-projection method was described in an alternative, more succinct form and then generalized to provide a robust means of back-projecting under projective transformations.

Experiments tested the invariant of two lines and two points, the invariant affine coordinates, the point-based moment invariants and the local moment invariants and concluded that the latter would most likely perform best in an actual classification system; it was also shown that affine back-projection based on two boundary points performs well.

The results presented above indicate that invariant features will have a valuable role to play in recognizing planar objects in the future, and that moment invariants are much more useful than is often assumed. Nevertheless there is scope for a considerable amount of further work; in particular, how the invariants tested in chapter 6 perform in an actual recognition system. It would also be interesting to see whether the projective differential invariants are robust enough for use in such a system.

Appendix A

Orthogonality of Rotation Invariants for Discrete Images

This appendix investigates the form of the matrix $\mathbf{V}^H\mathbf{V}$ as discussed in section 2.10.2. Consider the orthogonal polynomials of chapter 2, and let them be represented by $v_{pq}(x,y)$, p, $q = 1, 2, \ldots$. The moments z_{pq} of an image $f(x,y)$ are formed as follows:

$$z_{pq} = \int_{-\infty}^{+\infty} \int_{-\infty}^{+\infty} v_{pq}^*(x,y) f(x,y)\, dx\, dy, \qquad (A.1)$$

where v^* is the complex-conjugate of v. If we replace the integrals by summations and adopt the notation of chapter 2, we can see that (A.1) becomes $\mathbf{z} = \mathbf{V}^H\mathbf{g}$. In other words, $\mathbf{V}^H$ is used instead of $\mathbf{V}^{-1}$. The condition of orthogonality in the discrete case becomes $\mathbf{V}^H\mathbf{V} = \mathbf{I}$, where $\mathbf{I}$ is the $M \times M$ identity matrix, and M is the number of points in the image (in the previous section M was equal to N^2; if for example we use a square grid of points defined over the unit circle, $M \neq N^2$). We will see below that $\mathbf{V}^H\mathbf{V} \neq \mathbf{I}$; in fact, if one has P radial functions, it is of the form

$$\mathbf{V}^H\mathbf{V} = \begin{bmatrix} \mathbf{A} & 0 & 0 & 0 & \mathbf{B_1} & 0 & 0 & 0 & \mathbf{B_2} & 0 & \cdots \\ 0 & \mathbf{A} & 0 & 0 & 0 & \mathbf{B_1} & 0 & 0 & 0 & \mathbf{B_2} & \cdots \\ 0 & 0 & \mathbf{A} & 0 & 0 & 0 & \mathbf{B_1} & 0 & 0 & 0 & \cdots \\ 0 & 0 & 0 & \mathbf{A} & 0 & 0 & 0 & \mathbf{B_1} & 0 & 0 & \cdots \\ \mathbf{B_1} & 0 & 0 & 0 & \mathbf{A} & 0 & 0 & 0 & \mathbf{B_1} & 0 & \cdots \\ \vdots & \vdots & \vdots & \vdots & \vdots & \vdots & \vdots & \vdots & \vdots & \vdots \end{bmatrix} \quad \begin{array}{l} \mathbf{A}, \mathbf{B}_i \text{ are} \\ P \times P \text{ matrices.} \end{array}$$

The orthogonal moments of chapter 2 use polar separable functions of the form $h_p(r)e^{jq\theta}$. Let us assume we have P radial functions and Q angular ones such that $PQ = M$; also let $k = p + Q*(q-1)$, $k' = p' + Q*(q'-1)$, let the element in row k' and column k of $\mathbf{V}^H\mathbf{V}$ be denoted by $(\mathbf{V}^H\mathbf{V})_{(p'q')(pq)}$, and let (r_i, θ_i), $1 \leq i \leq M$, be the polar coordinates of all points on the image, then

$$(\mathbf{V}^H\mathbf{V})_{(p'q')(pq)} = \sum_i r_i h_{p'}^*(r_i) h_p(r_i) e^{j(q-q')\theta_i}. \qquad (A.2)$$

Note the factor r_i which comes from changing cartesian coordinates to polar coordinates when integrating. To be orthogonal, we require

$$(\mathbf{V}^H\mathbf{V})_{(p'q')(pq)} = \delta_{p'p}\delta_{q'q}, \quad \text{with } \delta_{ij} \text{ the Kronecker delta function.}$$

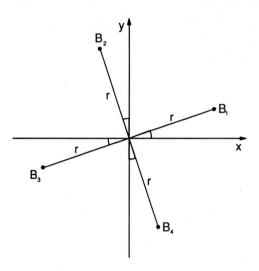

Figure A.1: A square grid of points has 4-fold rotational symmetry.

Now, as depicted in Fig. A.1, an $N \times N$ square sampling grid has fourfold rotational symmetry — every point on the grid has three other points with equal radius at angles of 90^0, 180^0 and 270^0 to the first point. This means that there are four cases to look at when considering the form of $\mathbf{V}^H\mathbf{V}$:

1. $q - q' = 0$.
2. $q - q'$ odd.
3. $q - q'$ even, but not divisible by 4.
4. $q - q'$ divisible by 4.

In the following we will let $s = q - q'$.

Case 1
(A.3) tells us that, when $s = 0$,

$$a_{p'p} \triangleq (\mathbf{V}^H\mathbf{V})_{(p'q')(pq)} = \sum_i r_i h_{p'}^*(r_i)h_p(r_i).$$

Even if the continuous functions $h_p(r)$ are orthogonal, the discrete version of the inner product gives values that are not precisely orthogonal i.e. $a_{p'p} \neq \delta_{p'p}$.

Case 2
If s is odd, we can write it as $s = 2k + 1$, k integer. If we take all opposed pairs of image points, such as B_1 and B_3 in Fig. A.1, we can write (A.3) as

$$\sum_{i \text{ over all pairs}} h_{p'}^*(r_i)h_p(r_i) \left[e^{js\theta_i} + e^{js(\theta_i + \pi)} \right]. \tag{A.3}$$

Now, $s = 2k + 1$, $e^{j2k\pi} = 1$ when k is an integer, and $e^{j\pi} = -1$, so

$$e^{js(\theta_i + \pi)} = e^{(2k+1)(\theta_i + \pi)} = e^{j2k\pi} e^{j\pi} e^{j(2k+1)\theta_i} = -e^{js\theta_i},$$

from which we see that the expression in (A.6) is zero:

$$(\mathbf{V}^H\mathbf{V})_{(p'q')(pq)} = 0 \quad \text{if } q - q' \text{ is odd.}$$

Case 3

If s is even but not divisible by 4 we can write it as $s = 4k + 2$, k integer. If we take all pairs of points separated by $90°$, such as B_1 and B_2 in Fig. A.1, (A.3) becomes

$$\sum_{i \text{ over all pairs}} h_{p'}^*(r_i) h_p(r_i) \left[e^{js\theta_i} + e^{js(\theta_i + \pi/2)} \right].$$

But when $s = 4k + 2$, $e^{js\pi/2} = -1$ and the above expression is zero:

$$(\mathbf{V}^H\mathbf{V})_{(p'q')(pq)} = 0 \quad \text{if } q - q' \text{ is even but not divisible by 4.}$$

Case 4

When $s = 4k$, k integer, the fourfold symmetry of the sampling grid means that each point in the quartet of Fig. A.1 maps to the same point: in general

$$e^{j4k\theta_i} = e^{j4k(\theta_i + \pi/2)} = e^{j4k(\theta_i + \pi)} = e^{j4k(\theta_i + 3\pi/2)} \neq 0.$$

Hence

$$(B_{s/4})_{p'p} \triangleq (\mathbf{V}^H\mathbf{V})_{(p'q')(pq)} \neq 0 \quad \text{when } q - q' \text{ divisible by 4.}$$

Summary

If we let the kth column of $\mathbf{V}$ be the M-dimensional vector $\mathbf{v}_{pq}$, $k = p + Q * (q - 1)$, then

$$\mathbf{V} = [\mathbf{v}_{11}, \mathbf{v}_{21}, \ldots, \mathbf{v}_{P1}, \mathbf{v}_{12}, \ldots, \mathbf{v}_{P2}, \ldots, \mathbf{v}_{1Q}, \ldots, \mathbf{v}_{PQ}].$$

If we define $\mathbf{V}_i$ as the $M \times P$ matrix with columns $\mathbf{v}_{pi}$, $1 \leq p \leq P$, then

$$\mathbf{V} = [\mathbf{V}_1 \ \mathbf{V}_2 \ \cdots \ \mathbf{V}_Q],$$

and

$$\mathbf{V}^H\mathbf{V} = \begin{bmatrix} \mathbf{V}_1^H\mathbf{V}_1 & \mathbf{V}_1^H\mathbf{V}_2 & \cdots & \mathbf{V}_1^H\mathbf{V}_Q \\ \mathbf{V}_2^H\mathbf{V}_1 & \mathbf{V}_2^H\mathbf{V}_2 & \cdots & \\ \vdots & \vdots & & \\ \mathbf{V}_Q^H\mathbf{V}_1 & \cdots & & \mathbf{V}_Q^H\mathbf{V}_Q \end{bmatrix}.$$

The above results tell us that $\mathbf{V}_{q'}^H\mathbf{V}_q = \mathbf{0}$ unless $q = q'$ or $q - q'$ is divisible by 4. Hence

$$\mathbf{V}^H\mathbf{V} = \begin{bmatrix} \mathbf{A} & \mathbf{0} & \mathbf{0} & \mathbf{0} & \mathbf{B}_1 & \mathbf{0} & \mathbf{0} & \mathbf{0} & \mathbf{B}_2 & \mathbf{0} & \cdots \\ \mathbf{0} & \mathbf{A} & \mathbf{0} & \mathbf{0} & \mathbf{0} & \mathbf{B}_1 & \mathbf{0} & \mathbf{0} & \mathbf{0} & \mathbf{B}_2 & \cdots \\ \mathbf{0} & \mathbf{0} & \mathbf{A} & \mathbf{0} & \mathbf{0} & \mathbf{0} & \mathbf{B}_1 & \mathbf{0} & \mathbf{0} & \mathbf{0} & \cdots \\ \mathbf{0} & \mathbf{0} & \mathbf{0} & \mathbf{A} & \mathbf{0} & \mathbf{0} & \mathbf{0} & \mathbf{B}_1 & \mathbf{0} & \mathbf{0} & \cdots \\ \mathbf{B}_1 & \mathbf{0} & \mathbf{0} & \mathbf{0} & \mathbf{A} & \mathbf{0} & \mathbf{0} & \mathbf{0} & \mathbf{B}_1 & \mathbf{0} & \cdots \\ \vdots & \vdots & \vdots & \vdots & \vdots & \vdots & \vdots & \vdots & \vdots & \vdots & \end{bmatrix} \quad \begin{array}{l} \mathbf{A}, \mathbf{B}_i \text{ are} \\ P \times P \text{ matrices.} \end{array}$$

with

$$(\mathbf{A})_{p'p} = \sum_i r_i h_{p'}^* h_p(r_i);$$

$$(\mathbf{B}_k)_{p'p} = \sum_i r_i h_{p'}^* h_p(r_i) e^{j4k\theta_i},$$

as stated in chapter 2.

Appendix B

Generating Algebraic and Moment Invariants

This appendix explains how to go about finding algebraic invariants. The most transparent technique is to use tensor contraction, which is a notationally succinct way of performing the operations introduced in the nineteenth century. To generate invariants of higher order forms by hand, one should resort to the historic techniques; however, the tensor-based method is amenable to implementation on a computer, and thus allows one to generate invariants relatively painlessly. Salmon lists a large number of invariants [61], which are repeated in section 3.4; the author has generated two further invariants using the historic techniques, and one invariant on a computer using the tensor method. Since the historic techniques give more of an insight into the meaning of invariants, they are summarized in the first section below; it may be omitted on a first reading, since the techniques are superceded by the tensor method, which is conceptually simpler and more concise, and is described in section B.2. Recently, Taubin [77] has described an alternative method of deriving algebraic invariants, but it is not simpler than using tensors.

The penultimate section gives a brief discussion of Cayley and Sylvester's approach to computing the number of irreducible invariants of a given form or system of forms, while the last section gives a proof of the revised fundamental theorem of moment invariants (RFTMI) and shows how to obtain moment invariants directly using tensors.

B.1 The historic method

First, let us define resultants, discriminants and covariants. These are important because

a) The *discriminant* of a form, or the *resultant* of a system of forms, is an invariant.

b) *Covariants* are derived from a form or a system of forms and have the property that any invariant of a covariant or a system of covariants is an invariant of the original form(s).

c) All invariants can be found using (a) and (b).

B.1.1   Resultants

Given a number of forms, the resultant is the function of the coefficients that is zero when the forms all share a common root. As an example, the resultant of $at + b = 0$ and $At + B = 0$ is $aB - bA$, which we came across earlier.

If we have two polynomials

$$
\begin{align}
\psi(t) &= t^m + p_1 t^{m-1} + \ldots = 0, &\text{(B.1)} \\
\phi(t) &= t^n + q_1 t^{n-1} + \ldots = 0, &\text{(B.2)}
\end{align}
$$

with roots $\{\alpha_1, \beta_1, \ldots\}$ and $\{\alpha_2, \beta_2, \ldots\}$ respectively, then we can represent (B.1) and (B.2) as

$$
\begin{align}
\psi(t) &= (t - \alpha_1)(t - \beta_1)(t - \gamma_1)\cdots = 0; \\
\phi(t) &= (t - \alpha_2)(t - \beta_2)(t - \gamma_2)\cdots = 0. &\text{(B.3)}
\end{align}
$$

If the second polynomial is to have a common root with the first then clearly one of the functions $\phi(\alpha_1)$, $\phi(\beta_1)$, $\phi(\gamma_1)$ etc. must be zero, so the product of all these functions $\phi(\alpha_1)\phi(\beta_1)\phi(\gamma_1)\cdots$ must be zero. By looking at (B.3) we see that the resultant is the product of all possible differences between a root of the first equation and a root of the second:

$$
\phi(\alpha_1)\phi(\beta_1)\cdots = (\alpha_1 - \alpha_2)(\alpha_1 - \beta_2)\cdots(\beta_1 - \alpha_2)(\beta_1 - \beta_2)\cdots = 0. \qquad \text{(B.4)}
$$

In the case of homogeneous polynomials $t = x/y$, and we can write the roots of the first equation as $\alpha_1 = x_1/y_1$, $\beta_1 = x_2/y_2$ etc. and those of the second equation as $\alpha_2 = x_1'/y_1'$, $\beta_2 = x_2'/y_2'$ etc. (B.3) then becomes

$$
(xy_1' - x_1'y)(xy_2' - x_2'y)\cdots(xy_m' - x_m'y) = 0.
$$

Similarly the resultant (B.4) becomes

$$
(x_1 y_1' - x_1' y_1)(x_1 y_2' - x_2' y_1)\cdots(x_2 y_1' - x_1' y_2)(x_2 y_2' - x_2' y_2)\cdots = 0.
$$

Resultants are easy to compute; a number of ways of obtaining them are described by Salmon [61]. The easiest to understand, if not the most efficient to apply, is Sylvester's dialytic method [133]. To find the resultant of two binary forms $ax^m + mx^{m-1}y + \ldots$ (order m) and $a'x^n + nb'x^{n-1}y + \ldots$ (order n) multiply the former by $x^{n-1}, x^{n-2}y, \ldots, y^{n-1}$ and the latter by $x^{m-1}, x^{m-2}y, \ldots, y^{m-1}$. This gives us $m+n$ linear equations in the $m+n$ terms $x^{m+n-1}, x^{m+n-2}y, \ldots, y^{m+n-1}$. If we consider the latter as unknowns we can eliminate them, leaving us with the resultant. The simplest case of such elimination is finding the resultant of two first order equations

$$
\begin{align}
ax &+ by &= 0, \\
Ax &+ By &= 0.
\end{align}
$$

The resultant is: $\quad \begin{vmatrix} a & b \\ A & B \end{vmatrix} = aB - bA.$

Sylvester's method applied to the two quadratics $ax^2 + 2bxy + cy^2 = 0$ and $Ax^2 + 2Bxy + Cy^2$ requires multiplication of each by x and by y, giving the four equations

$$
\begin{array}{rcl}
ax^3 + 2bx^2y + cxy^2 & = & 0, \\
ax^2y + 2bxy^2 + cy^3 & = & 0, \\
Ax^3 + 2Bx^2y + Cxy^2 & = & 0, \\
Ax^2y + 2Bxy^2 + Cy^3 & = & 0.
\end{array}
$$

Eliminating x^3, x^2y, xy^2 and y^3 gives us the resultant in the form of a determinant:

$$
\text{Resultant} \quad = \quad
\begin{vmatrix}
a & 2b & c & 0 \\
0 & a & 2b & c \\
A & 2B & C & 0 \\
0 & A & 2B & C
\end{vmatrix}
$$

$$
= \quad (aC - cA)^2 + 4(bA - aB)(bC - cB).
$$

Resultants are homogeneous functions of the coefficients of each equation, and for equations of the mth and nth order respectively, the resultant is of the nth order in the coefficients of the first equation and of the mth order in the coefficients of the second.

B.1.2 Discriminants

The discriminant of a form in k variables is obtained by differentiating it with respect to each variable and finding the resultant of the k differentials. The discriminant is equal to the product of the squares of all the differences of any two roots of the form; if the roots are $\{x_iy_i\}$, then it equals

$$
(x_1y_2 - y_1x_2)^2(x_1y_3 - y_1x_3)^2 \cdots .
$$

As an example, the discriminant of $ax^2 + 2bxy + cy^2$ is $ac - b^2$.

B.1.3 Invariants

Theorem: The discriminant of a form or the resultant of a system of forms is an invariant.

Proof: Let U be a binary form, and (x_1, y_1) be a root of U; then $xy_1 - yx_1$ is a factor of U. If we transform the variables, so that $x = \alpha x' + \beta y'$ and $y = \gamma x' + \delta y'$, then the factor $xy_1 - yx_1$ becomes $(\alpha x' + \beta y')y_1 - (\gamma x' + \delta y')x_1$. If we write this as $x'Y_1 - y'X_1$ we have

$$
X_1 = -(\beta y_1 - \delta x_1); \qquad Y_1 = \alpha y_1 - \gamma x_1.
$$

We will get corresponding expressions for the other X_i and Y_i. Now, one can easily show that

$$
X_iY_j - X_jY_i = \Delta(x_iy_j - x_jy_i). \tag{B.5}
$$

with $\Delta = (\alpha\delta - \beta\gamma)$. In other words, $x_iy_j - x_jy_i$ is an invariant of weight 1. Since the resultant and the discriminant are equal to the product of a number of terms $x_iy_j - x_jy_i$, it is clear that the discriminant or resultant of the transformed system is equal to the original multiplied by a power of Δ.

B.1.4 Covariants

A covariant ϕ is defined in exactly the same way as an invariant I, only a covariant is also a function of the variables x, y. If a linear transformation of the coordinates results in

$$(a_0', a_1', \ldots, a_p')(x', y')^p = (a_0, a_1, \ldots, a_p)(x, y)^p \tag{B.6}$$

then ϕ is a covariant of the form if

$$\phi(a_0', a_1', \ldots, a_p', x', y') = \Delta^g \phi(a_0, a_1, a, \ldots, a_p, x, y).$$

Clearly, the original function (B.6) is itself a covariant with $w = 0$. Finding covariants will be discussed below; for now, here is an example of a covariant of any binary form U of order greater than 2 (it will be an invariant if U is of order 2):

$$\text{A covariant of } U: \qquad \frac{\partial^2 U}{\partial x^2} \frac{\partial^2 U}{\partial y^2} - \left(\frac{\partial^2 U}{\partial x \partial y} \right)^2.$$

B.1.5 Finding covariants and invariants

Having defined resultants, discriminants and covariants, we are now in a position to see how to go about finding covariants and hence invariants. To this end I have summarised below the essentials of Cayley's symbolic representation, introduced in 1846 [134].

First, we need to see how the partial differentials $\partial/\partial x$ and $\partial/\partial y$ are affected by a linear transformation. With $\Delta = \alpha\delta - \beta\gamma$, let

$$
\begin{array}{llll}
x & = & \alpha X + \beta Y; & \quad X & = & \frac{1}{\Delta}(\delta x - \beta y); \\
y & = & \gamma X + \delta Y; & \quad Y & = & \frac{1}{\Delta}(-\gamma x + \alpha y).
\end{array}
\tag{B.7}
$$

If we know that the form $f_p(x,y)$ becomes $f_p'(X,Y)$, what can we say about $\partial f_p/\partial x$ and $\partial f_p/\partial y$ in terms of $\partial f_p'/\partial X$ and $\partial f_p'/\partial Y$? Well, we can write

$$\frac{\partial}{\partial x} = \frac{\partial}{\partial X}\frac{\partial X}{\partial x} + \frac{\partial}{\partial Y}\frac{\partial Y}{\partial x} = \frac{1}{\Delta}\left(\delta\frac{\partial}{\partial X} - \gamma\frac{\partial}{\partial Y} \right),$$

$$\frac{\partial}{\partial y} = \frac{\partial}{\partial X}\frac{\partial X}{\partial y} + \frac{\partial}{\partial Y}\frac{\partial Y}{\partial y} = \frac{1}{\Delta}\left(-\beta\frac{\partial}{\partial X} + \alpha\frac{\partial}{\partial Y} \right),$$

which can be written as

$$
\begin{bmatrix} \partial/\partial y \\ -\partial/\partial x \end{bmatrix} = \frac{1}{\Delta} \begin{bmatrix} \alpha & \beta \\ \gamma & \delta \end{bmatrix} \begin{bmatrix} \partial/\partial Y \\ -\partial/\partial X \end{bmatrix}.
$$

Comparing this with (B.7) indicates that, except for the factor $1/\Delta$, $\partial/\partial y$ and $-\partial/\partial x$ are transformed in the same way that x and y are. In particular, if $f_p(x,y) = f_p'(X,Y)$, then

$$f_p\left(\frac{\partial}{\partial y}, -\frac{\partial}{\partial x} \right) = \frac{1}{\Delta^p} f_p'\left(\frac{\partial}{\partial Y}, -\frac{\partial}{\partial X} \right). \tag{B.8}$$

Although this presents a way of finding invariants and covariants, we are interested in the result for a different reason. Earlier we saw that $x_1y_2 - x_2y_1$ is an invariant (B.5). Using the above result allows us to say that

$$\overline{12} = \frac{\partial}{\partial x_1}\frac{\partial}{\partial y_2} - \frac{\partial}{\partial x_2}\frac{\partial}{\partial y_1}$$

is an invariant operator, where $\overline{12}$ is the symbolic notation. We can use this to form covariants or invariants of two forms U and V by writing the variables of U with the subscript '1' and the variables of V with the subscript '2', operating on the product UV with any power of $\overline{12}$ and then dropping the subscripts. If after the operation the variables x and y are no longer present, we have obtained an invariant; otherwise we have a covariant. $\overline{12}\{UV\}$ gives the Jacobian

$$J = \frac{\partial U}{\partial x}\frac{\partial V}{\partial y} - \frac{\partial U}{\partial y}\frac{\partial V}{\partial x},$$

and

$$\overline{12}^2\{UV\} = \frac{\partial^2 U}{\partial x^2}\frac{\partial^2 V}{\partial y^2} + \frac{\partial^2 U}{\partial y^2}\frac{\partial^2 V}{\partial x^2} - 2\frac{\partial^2 U}{\partial x\partial y}\frac{\partial^2 V}{\partial x\partial y}.$$

When

$$U = a_1x_1^2 + 2b_1x_1y_1 + c_1y_1^2, \qquad V = a_2x_2^2 + 2b_2x_2y_2 + c_2y_2^2$$

then

$$\overline{12}^2\{UV\} = a_1c_2 + a_2c_1 - 2b_1b_2,$$

the so-called intermediate invariant, discovered by Boole in 1841.

The above method can also be used to obtain invariants or covariants of a single form U, simply by setting $V = U$. Hence $\frac{1}{2}\overline{12}^2\{U\}$ becomes the covariant known as the Hessian

$$\frac{\partial^2 U}{\partial x^2}\frac{\partial^2 U}{\partial y^2} - \left(\frac{\partial^2 U}{\partial x\partial y}\right)^2.$$

When $U = ax^2 + 2bxy + cy^2$, the above becomes the invariant $ac - b^2$.

Only even powers of $\overline{12}$ can be used on a single form, since $\overline{12}^n\{U\}$ is always zero when n is odd. As we have seen, $\overline{12}^2$ applied to a quadratic gives an invariant; similarly, $\overline{12}^4$ applied to the quartic $ax^4 + 4bx^3y + \ldots + ey^4$ yields the invariant $ae - 4bd + 3c^2$, and in general applying $\overline{12}^n$ to the binary form $(a_0, \ldots, a_n)(x, y)^n$ gives the invariant

$$a_0a_n - \binom{n}{1}a_1a_{n-1} + \binom{n-1}{2}a_2a_{n-2} + \ldots + \frac{1}{2}\binom{n}{\frac{n}{2}}a_{n/2}^2.$$

The above technique can be extended to any number of forms U, V, W etc. by giving the variables of U the subscript '1', those of V '2', W '3' etc. We can operate on their product with any number of symbols $\overline{12}^p$, $\overline{23}^q$, $\overline{31}^r$ etc. (p, q, r integer). As before one can find covariants of a single form U by forming $UVW \ldots = U_1U_2U_3 \ldots$ where U_i is the form U whose variables have the subscript i.

As an example, we can investigate the possible invariants of order 3 in the coefficients. Such invariants are formed by operating on $U_1U_2U_3$, so the operator is of the form $\overline{12}^p.\overline{23}^q.\overline{31}^r$. In order to yield an invariant the power of $\partial/\partial x_i$ and $\partial/\partial y_i$ in the operator must be equal to n, i.e.

$$p + r = p + q = q + r = n \qquad \Rightarrow p = q = r.$$

The general form of the operator is then $(\overline{12}.\overline{23}.\overline{31})^\alpha$; as before, the derivative is identically zero if α is odd, so all invariants of the third order in the coefficients are given by using the operator $(\overline{12}.\overline{23}.\overline{31})^\alpha$ with α even.

As an example, setting $\alpha = 2$ and applying the operator to the quartic with coefficients a, b, c, d, e gives the invariant

$$ace + 2bcd - ad^2 - eb^2 - c^3.$$

The interested reader is referred to the books by Salmon [61] and Clebsch [69] for a more thorough exposition.

B.2 The tensor method

This section shows how to generate algebraic invariants of forms in any number of variables using tensors; it is a summary of the techniques presented by Gurevich [64]. The operations performed when using tensors are the same as those using Aronhold's symbolic method [69, 64, 9], but tensors are preferable because their operation is more transparent. As we will see, their use is a further development of Cayley's symbolic method.

An important advantage of the tensor method is that it is symbolic, and hence can be implemented on a computer; the author has written a program that generates invariants, and has used it to verify some of the invariants listed by Salmon as well as to generate a new invariant (see section 3.4).

B.2.1 Contravariant and covariant tensors

A *contravariant* n-dimensional vector $\mathbf{x}$ is written as x^i in tensor notation, with elements x^1, x^2, ..., x^n. A linear transformation in n-dimensions is defined by p_α^i, $\alpha, i = 1, 2, \ldots, n$ (an $n \times n$ matrix with elements $p_1^1, p_2^1, \ldots, p_1^2, p_2^2$ etc.), and transforms the vector x^i to $\hat{x}^i$ as follows[1] [64]:

$$x^i = \sum_{\alpha=1}^{n} p_\alpha^i \hat{x}^\alpha, \qquad i = 1, 2, \ldots, n.$$

Under the Einstein summation convention we can write the above as

$$x^i = p_\alpha^i \hat{x}^\alpha, \qquad i = 1, 2, \ldots, n, \tag{B.9}$$

since a pair of identical sub- and superscripts signifies summation over all their values. In the following we will assume that the determinant of the transformation is non-zero:

$$\Delta = |p_\alpha^i| \neq 0.$$

For the 2-D case, $n = 2$ and

$$\Delta = \begin{vmatrix} p_1^1 & p_2^1 \\ p_1^2 & p_2^2 \end{vmatrix} = p_1^1 p_2^2 - p_1^2 p_2^1.$$

---

[1]Cyganski & Orr [58, 78] use a different definition, in which x and $\hat{x}$ are interchanged.

A *covariant* n-dimensional vector is written as u_i and is transformed by the linear transformation p_i^α according to

$$\hat{u}_\alpha = p_\alpha^i u_i, \qquad \alpha = 1, 2, \ldots, n. \tag{B.10}$$

If one thinks of a contravariant vector as a column vector, then a covariant vector is a row vector. It follows from (B.9) and (B.10) that $u_i x^i = \hat{u}_\alpha \hat{x}^\alpha$. A covariant tensor of *order r* is written as $a_{i_1 i_2 \cdots i_r}$ and is transformed by the linear transformation p_α^i according to

$$\hat{a}_{\alpha_1 \alpha_2 \cdots \alpha_r} = p_{\alpha_1}^{i_1} p_{\alpha_2}^{i_2} \cdots p_{\alpha_r}^{i_r} a_{i_1 i_2 \cdots i_r}, \qquad i_j, \alpha_j = 1, 2, \ldots, n, \text{ for } 1 \le j \le r.$$

B.2.2 Symmetric and skew-symmetric tensors

A contravariant tensor is *symmetric* if its value is unchanged by swapping any number of indices e.g. a^{ijk} is symmetric if $a^{ijk} = a^{jik} = a^{jki} = a^{ikj} = a^{kij} = a^{kji}$ for all values of i, j and k. As an example of a symmetric tensor, consider the binary form $ax^2 + 2bxy + cy^2$. If we let $x_1 = x$ and $x_2 = y$, then this is $ax_1^2 + 2bx_1x_2 + cx_2^2$. If we let a^{ij} be a symmetric contravariant tensor, then we can express this as

$$ax_1^2 + 2bx_1x_2 + cx_2^2 = a^{ij}x_i x_j \tag{B.11}$$

by defining $a^{11} = a$; $a^{21} = a^{12} = b$; $a^{22} = c$.

Similarly, a general binary form can be expressed as a tensor product:

$$a_0 x^n + na_1 x^{n-1} y + \ldots + a_n y^n = a^{i_1 i_2 \cdots i_n} x_{i_1} x_{i_2} \cdots x_{i_n} \tag{B.12}$$

by defining $a^{i_1 i_2 \cdots i_n} = a_k$ if k of the superscripts $i_1, i_2, \ldots, i_n$ are equal to 1 and the remaining $n - k$ are equal to 2. $a^{i_1 i_2 \cdots i_n}$ is hence a symmetric tensor.

A covariant or contravariant tensor is *skew-symmetric* with respect to two indices if interchanging the two indices only results in a change of sign. For example, a^{ijk} is skew-symmetric with respect to the first and third indices if $a^{ijk} = -a^{kji}$ for all values of i, j and k. When $i = k$, $a^{ijk} = 0$.

A covariant or contravariant tensor is a *polyvector* if it is skew-symmetric over all indices. Polyvectors of order two and three are called bivectors and trivectors respectively. In the following we will be interested in the unit bivector ϵ_{ij} in 2D defined by $\epsilon_{12} = 1$ and the unit trivector ϵ_{ijk} defined by $\epsilon_{123} = 1$.

B.2.3 Multiplying tensors

Multiplying tensors is straightforward: given two contravariant tensors a^{ij}, b^{klm}, we define their product as the 5th order tensor

$$c^{ijklm} = a^{ij} b^{klm}.$$

Multiplying a covariant tensor by a contravariant tensor proceeds in the same way, except that the summation convention means that the sum is taken over all identical pairs of sub- and superscripts. For example, multiplying a^{ij} by d_{ikl} results in

$$e_{kl}^j = a^{ij} d_{ikl}.$$

B.2.4 Relative and absolute tensors

A contravariant tensor is defined as being a *relative tensor* of *weight g* if

$$a^{i_1 \cdots i_r} = \Delta^g p^{i_1}_{\alpha_1} \cdots p^{i_r}_{\alpha_r} \hat{a}^{\alpha_1 \cdots \alpha_r},$$

where $\Delta = |p^i_\alpha|$ is the determinant of the linear transformation. When $g = 0$ the tensor is called an *absolute tensor*. A special case is when the tensor is a scalar — it is then called a *relative invariant* if $g \neq 0$ and an *absolute invariant* if $g = 0$. An invariant σ of weight g becomes $\hat{\sigma}$ under linear transformation, where

$$\sigma = \Delta^g \hat{\sigma}.$$

A covariant relative tensor of weight g is defined as

$$\Delta^g \hat{a}_{\alpha_1 \cdots \alpha_r} = p^{i_1}_{\alpha_1} \cdots p^{i_r}_{\alpha_r} a_{i_1 \cdots i_r}. \qquad (B.13)$$

The weights of tensors are added under multiplication. Thus, if $x^i = \Delta^g p^i_\alpha x^\alpha$ and $\Delta^w \hat{u}_i = p^\alpha_i u_\alpha$, then $x^i u_i = \Delta^{g+w} \hat{x}^\alpha \hat{u}_\alpha$; i.e. $x^i u_i$ is a relative invariant of weight $g + w$.

The aim as far as image invariants are concerned is to find absolute invariants by using the ratio of powers of relative invariants that are constructed from the coefficients of binary forms [27].

We will now consider an example of an absolute tensor and of a relative tensor. The coefficients of a binary form are defined to have the following behaviour under linear transformations:

$$\hat{a}^{\alpha_1 \cdots \alpha_r} \hat{x}_{\alpha_1} \cdots \hat{x}_{\alpha_r} = a^{i_1 \cdots i_r} x_{i_1} \cdots x_{i_r}.$$

Using $\hat{x}_i = p^\alpha_i x_\alpha$ it is easy to see that the symmetric tensor $a_{i_1 \cdots i_r}$ used to define a binary form is an absolute tensor. On the other hand, the covariant unit polyvector can be shown to be a relative tensor of weight $g = 1$ if we stipulate that the transformed unit polyvector is also a unit polyvector, i.e. $\hat{\epsilon}_{i_1 \cdots i_r} = \epsilon_{i_1 \cdots i_r}$. For example, in 2-D we can show that

$$\Delta \hat{\epsilon}_{\alpha\beta} = p^i_\alpha p^j_\beta \epsilon_{ij},$$

by writing out $f_{\alpha\beta} = p^i_\alpha p^j_\beta \epsilon_{ij}$ in full:

$$f_{11} = p^1_1 p^2_1 - p^2_1 p^1_1 = 0; \qquad f_{12} = p^1_1 p^2_2 - p^2_1 p^1_2 = \Delta;$$

$$f_{21} = p^1_2 p^2_1 - p^1_1 p^2_2 = -\Delta; \qquad f_{22} = p^1_2 p^2_2 - p^2_2 p^1_2 = 0.$$

Hence we see from (B.13) that ϵ_{ij} has weight $g = 1$.

B.2.5 Total alternation produces invariants

Multiplying a relative contravariant tensor $a^{i_1 \cdots i_r}$ of weight g by $\epsilon_{i_1 \cdots i_n}$, $1 \leq i_j \leq r$, is known as total alternation with respect to the n indices $\{i_j\}$, and results in a relative tensor of weight $g + 1$ and order $r - n$. The result of total alternation is zero if $n > r$, or if $a^{i_1 \cdots i_r}$ is symmetric with respect to any of the n indices. In 2D ($n = 2$), if r is even ($r = 2q$, q integer) we can perform the above on any q pairs of indices, resulting in a relative invariant of weight $g + q$. This is known as *total alternation* of the tensor $a^{i_1 \cdots i_r}$. In n dimensions, we require $r = nq$ to be able to perform total alternation, using q sets of n indices resulting in an invariant of weight $g + q$ as before.

B.2.6 How to find an invariant of a given order in the coefficients of various binary forms

The tensor $t^{ijkl} = a^{ij}a^{kl}$ contains all possible 2nd order products of the coefficients of the quadratic binary form $a^{ij}x_ix_j$. Generalizing this, we can multiply q_1 tensors a^{ij} with q_2 tensors b^{lmn}, q_3 tensors c^{pqrs}, ... and q_{k-1} tensors $c^{i_1\cdots i_k}$ to get a contravariant tensor $t^{i_1\cdots i_r}$ of order

$$r = 2q_1 + 3q_2 + \ldots + kq_{k-1}$$

which contains all possible products of the coefficients in which the coefficients of the quadratic form appear q_1 times, those of the cubic form q_2 times etc.

As a consequence of Theorem 17.4 (p.187) in Gurevich's book [64], all the invariants of the system of $k-1$ binary forms of order r in the coefficients as above can be obtained by total alternation of the product $t^{i_1\cdots i_r}$.

B.2.7 Examples of finding invariants

Total alternation of $a^{ij}a^{kl}$ in 2D, where a^{ij} is symmetric, results in

$$a^{ij}a^{kl}\epsilon_{ik}\epsilon_{jl} = a^{11}a^{22} - a^{12}a^{21} - a^{21}a^{12} + a^{22}a^{11} = 2(ac - b^2),$$

where the last step follows from (B.11). The weight of the invariant is $g = 2$ since ϵ_{ij} is used twice. Similarly, total alternation of $a^{ij}b^{klm}b^{nop}$ gives

$$a^{ij}b^{klm}b^{nop}\epsilon_{ik}\epsilon_{jn}\epsilon_{lo}\epsilon_{mp} = 2\left[A(bd - c^2) - B(ad - bc) + C(ac - b^2)\right],$$

with $A = a^{11}$ etc., $a = b^{111}$, $b = b^{112}$ etc., which has weight $g = 4$. Both these invariants appear in [26, 27]. Most of the possible total alternations will be identically equal to zero. In particular, multiplying a symmetric tensor $a^{i_1\cdots i_r}$ by $\epsilon_{i_k i_l}$ results in zero:

$$a^{i_1\cdots i_r}\epsilon_{i_k i_l} \equiv 0, \qquad 1 \le k, l \le r.$$

B.3 Computing the number of irreducible invariants

We saw in section 3.3 that Hilbert proved that there are a finite number of irreducible invariants. One can in fact calculate the number of invariants a given form or system of forms will have of a given order in the coefficients, based on a method introduced by Cayley [71] and formally proven by Sylvester [72]. When dealing with an invariant of order θ in the coefficients of a single binary form of order n, the weight g of the invariant is given by $g = \frac{1}{2}n\theta$. If we let $p(g, n, \theta)$ be the number of ways of choosing θ integers from the set $\{0, 1, 2, \ldots, n\}$ so that their sum equals g, repetitions allowed, then the number of invariants is given by $N_i = p(g, n, \theta) - p(g - 1, n, \theta)$. The number of irreducible invariants of order θ can easily be found from this if we know the number of irreducible invariants of all orders less than θ; using these we can compute the number N_c of invariants of order θ that are combinations of those of order less than θ, which allows us to say that the number of irreducible invariants of order θ is given by $N = N_i - N_c$. If we have a system of forms we replace $p(g, n, \theta)$

with $p(g, n_1, \theta_1, n_2, \theta_2, \ldots)$, where the weight g is now $g = \frac{1}{2}(n_1\theta_1 + n_2\theta_2 + \ldots)$ and $p(.)$ is the number of ways of choosing θ_1 integers (repetitions allowed) from the set $\{0, 1, \ldots, n_1\}$, θ_2 from the set $\{0, 1, \ldots, n_2\}$ etc. such that they sum to g. Sylvester presents a rule for simplifying the above calculation for single forms in [135] and presents tables of the number and order of invariants for the first ten binary forms [136] and for the systems of two binary forms of the first four orders [137].

The author has used the above technique implemented on a computer to calculate the number and order of the irreducible invariants of the system of binary, cubic and quartic forms; they can be found in table 3.1.

B.4 Moment invariants using tensors

This final section proves the revised theorem of moment invariants (RFTMI) using the tensor methods discussed above, and gives an example of how to generate moment invariants directly using tensors.

B.4.1 The moment tensor

As we saw in chapter 2, the standard definition of the moment m_{pq} of an image $f(x, y)$ is

$$m_{pq} = \int_{-\infty}^{+\infty} \int_{-\infty}^{+\infty} x^p y^q f(x, y) \, dx \, dy. \tag{B.14}$$

If we let $x^1 = x$, $x^2 = y$, $dv = dx \, dy$ and $\mathbf{x} = (x, y)$, we can follow Cyganski & Orr [58] and define a symmetric moment tensor $M^{i_1 i_2 \cdots i_r}$, $r = p + q$, as

$$M^{i_1 \cdots i_r} = \int_V x^{i_1} x^{i_2} \cdots x^{i_r} f(\mathbf{x}) \, dv. \tag{B.15}$$

$M^{i_1 \cdots i_r} = m_{pq}$ if p of the superscripts are equal to 1 and q are equal to 2 — hence it can be seen to be a symmetric tensor. For an n-dimensional image, (B.15) still applies but the superscripts i_j can take on values between 1 and n, $f(\mathbf{x}) = f(x^1, x^2, \ldots, x^n)$ and $dv = dx^1 \, dx^2 \cdots dx^n$.

We can prove that $M^{i_1 \cdots i_r}$ is an oriented relative tensor of weight $g = 1$ as follows: first, let us define the transformed moment tensor $\hat{M}^{\alpha_1 \cdots \alpha_r}$ as

$$\hat{M}^{\alpha_1 \cdots \alpha_r} = \int_V \hat{x}^{\alpha_1} \hat{x}^{\alpha_2} \cdots \hat{x}^{\alpha_r} \, \hat{f}(\hat{\mathbf{x}}) \, d\hat{v},$$

where $d\hat{v} = d\hat{x}^1 \cdots d\hat{x}^n$ and $x^i = p_\alpha^i \hat{x}^\alpha$, with p_α^i being the linear transformation, and

$$f(\mathbf{x}) = f(x^1, \ldots, x^n) = f(p_\alpha^1 \hat{x}^\alpha, \ldots, p_\alpha^n \hat{x}^\alpha) = \hat{f}(\hat{x}^1, \ldots, \hat{x}^n) = \hat{f}(\hat{\mathbf{x}}).$$

To find the relationship between $M^{i_1 \cdots i_r}$ and $\hat{M}^{\alpha_1 \cdots \alpha_r}$ we insert $x^i = p_\alpha^i \hat{x}^\alpha$ into (B.15) giving

$$
\begin{aligned}
M^{i_1 \cdots i_r} &= \int_V p_{\alpha_1}^{i_1} \hat{x}^{\alpha_1} \cdots p_{\alpha_r}^{i_r} \hat{x}^{\alpha_r} \, \hat{f}(\hat{\mathbf{x}}) \, |J| \, d\hat{v} \\
&= |J| \, p_{\alpha_1}^{i_1} \cdots p_{\alpha_r}^{i_r} \int_V \hat{x}^{\alpha_1} \cdots \hat{x}^{\alpha_r} \, \hat{f}(\hat{\mathbf{x}}) \, d\hat{v} \\
&= |\Delta| \, p_{\alpha_1}^{i_1} \cdots p_{\alpha_r}^{i_r} \, \hat{M}^{\alpha_1 \cdots \alpha_r} \tag{B.16}
\end{aligned}
$$

where J is the Jacobian of the transformation, and $\Delta = J$ for linear transformations. Hence we see that the moment tensor has weight $g = 1$. $M^{i_1 \cdots i_r}$ is known as an oriented tensor [58] because $|\Delta|$ appears on the right hand side of (B.16) instead of Δ (a fact that Bamieh & de Figueiredo [76] miss).

B.4.2 The RFTMI using tensors

Comparing (B.16) with the relationship between the coefficients of a binary form under linear transformation

$$a^{i_1 \cdots i_r} = p^{i_1}_{\alpha_1} \cdots p^{i_r}_{\alpha_r} \hat{a}^{\alpha_1 \cdots \alpha_r} \tag{B.17}$$

allows one to prove the revised fundamental theorem of moment invariants [27] concisely as follows: an algebraic invariant of a binary form with coefficients $a^{11 \cdots 11}$, $a^{11 \cdots 12}, \ldots, a^{22 \cdots 22}$ is a homogeneous polynomial in these coefficients; assume a given invariant I is a homogeneous polynomial of order k i.e. each term in the polynomial is a product of k coefficients, repetitions allowed. Assume further that this invariant has weight g:

$$I(a^{11 \cdots 11}, a^{11 \cdots 12}, \ldots, a^{22 \cdots 22}) = \Delta^g I(\hat{a}^{11 \cdots 11}, \hat{a}^{11 \cdots 12}, \ldots, \hat{a}^{22 \cdots 22}).$$

From (B.16) and (B.17) we see that if we replace $a^{i_1 \cdots i_r}$ by the moment tensor $M^{i_1 \cdots i_r}$, we must replace $\hat{a}^{i_1 \cdots i_r}$ by $|\Delta| \hat{M}^{i_1 \cdots i_r}$, giving

$$
\begin{aligned}
I(M^{11 \cdots 11}, M^{11 \cdots 12}, \ldots, M^{22 \cdots 22}) &= \Delta^g I(|\Delta| \hat{M}^{11 \cdots 11}, |\Delta| \hat{M}^{11 \cdots 12}, \ldots, |\Delta| \hat{M}^{22 \cdots 22}) \\
&= \Delta^g |\Delta|^k I(\hat{M}^{11 \cdots 11}, \hat{M}^{11 \cdots 12}, \ldots, \hat{M}^{22 \cdots 22}),
\end{aligned}
$$

which proves the revised fundamental theorem of moment invariants as stated by the author in reference [27]. As an example, the second order invariant $I(a, b, c) = ac - b^2$ becomes, on replacing $\hat{a}$ by $|\Delta| \hat{m}_{20}$, $\hat{b}$ by $|\Delta| \hat{m}_{11}$ and $\hat{c}$ by $|\Delta| \hat{m}_{02}$:

$$
\begin{aligned}
I(m_{20}, m_{11}, m_{02}) &= \Delta^2 I(|\Delta| \hat{m}_{20}, |\Delta| \hat{m}_{11}, |\Delta| \hat{m}_{02}) = \Delta^2 \left\{ |\Delta|^2 \hat{m}_{20} \hat{m}_{02} - |\Delta|^2 \hat{m}_{11}^2 \right\} \\
&= \Delta^2 |\Delta|^2 I(\hat{m}_{20}, \hat{m}_{11}, \hat{m}_{02}).
\end{aligned}
$$

B.4.3 An example of a moment invariant generated using tensors

The feature that is invariant to linear image transformations and is the second order product of the second order image moments is given by (c.f. (4.3)):

$$
\begin{aligned}
I_1 = M^{ij} M^{kl} \epsilon_{ik} \epsilon_{jl} &= M^{11} M^{22} - M^{12} M^{21} - M^{21} M^{12} + M^{22} M^{11} \\
&= 2(m_{20} m_{02} - m_{11}^2),
\end{aligned}
$$

where $m_{20} = M^{11}$, $m_{11} = M^{12} = M^{21}$ and $m_{02} = M^{22}$. It has weight $g = 4$ since M^{ij} and ϵ_{ij} each have weight 1, and ϵ_{ij} appears an even number of times.

B.4.4   Moment invariants based on point features

When the image information is a set of points rather than the image intensity or binary shape, invariants are obtained by using the moment tensor of the feature points; this time, however, we will see that the moment tensor is of weight zero [78]. If we let $\{\mathbf{x}(k)\}$, $1 \leq k \leq N$, be a set of N feature points on the image with $\mathbf{x}(k) = (x^1(k), x^2(k))$, then we define the feature-point image $f(\mathbf{x})$ as

$$f(\mathbf{x}) = \sum_{\mathbf{k}} \delta(\mathbf{x} - \mathbf{x}(k)),$$

where $\delta(\mathbf{x})$ is the Dirac delta function. Using the definition of the moment tensor (B.15), we see that

$$
\begin{aligned}
M^{i_1 i_2 \cdots i_r} &= \int_V x^{i_1} x^{i_2} \cdots x^{i_r} f(\mathbf{x})\, dv \\
&= \int_V x^{i_1} x^{i_2} \cdots x^{i_r} \sum_{k} \delta(\mathbf{x} - \mathbf{x}(k))\, dv \\
&= \sum_{k} x^{i_1}(k) x^{i_2}(k) \cdots x^{i_r}(k).
\end{aligned}
$$

Putting $x^i = p^i_\alpha \hat{x}^i$ into this gives

$$
\begin{aligned}
M^{i_1 i_2 \cdots i_r} &= \sum_{k} p^{i_1}_{\alpha_1} \hat{x}^{\alpha_1}(k)\, p^{i_2}_{\alpha_2} \hat{x}^{\alpha_2}(k) \cdots p^{i_r}_{\alpha_r} \hat{x}^{\alpha_r}(k) \\
&= p^{i_1}_{\alpha_1} p^{i_2}_{\alpha_2} \cdots p^{i_r}_{\alpha_r} \hat{M}^{\alpha_1 \alpha_2 \cdots \alpha_r},
\end{aligned}
$$

from which we see that $M^{i_1 \cdots i_r}$ is indeed an absolute tensor.

   To obtain absolute invariants one can no longer divide by μ_{00} since it is a constant equal to the number of points. Section 4.4.2 of chapter 4 discusses how to generate absolute invariants.

Appendix C

Equations Relating Solid Motion in 3D to Affine Image Transformations

We saw in chapter 1 that, under weak perspective, motion of a planar object causes affine transformations of its image. If one places the coordinate origin at the image's centroid, one removes the effect of translation. The remaining four parameters of the linear transformation are related to the three rotation parameters (α, β, γ) and the scale parameter K of object rotations and scaling in 3-D as shown below.

Assume that a planar object in 3D is scaled by a factor K, rotated clockwise about the x-axis by an amount α, clockwise about the y-axis by an amount β and clockwise about the z-axis by an amount γ in that order, then the 2-D image of the transformed object $f(x', y')$ is related to the original image $f(x, y)$ by the linear coordinate transformation

$$x' = Ax + By; \qquad y' = Cx + Dy.$$

If one chooses the z-axis to be perpendicular to the image plane and pointing away from the camera, the x-axis pointing upwards and the y-axis to the right, and one considers the motion of the two points $(1, 0, 0)$ and $(0, 1, 0)$ under the above four transformations, it is reasonably straightforward to show that

$$\tan \gamma \;=\; \frac{C}{A}; \tag{C.1}$$

$$\cos \beta \;=\; \frac{A}{K \cos \gamma}; \tag{C.2}$$

$$\cos \alpha \;=\; \frac{AD - BC}{AK} \cos \gamma. \tag{C.3}$$

We see from (C.1) that, given A, B, C and D we can determine γ straight away. If we define

$$b = D \sin \gamma + B \cos \gamma; \qquad d = D \cos \gamma - B \sin \gamma;$$

$$p = \frac{b}{d}; \qquad a = \frac{AD - BC}{A} \cos \gamma; \qquad c = \frac{A}{\cos \gamma}; \tag{C.4}$$

then one can show that

$$K^2 = \frac{1}{2}(a^2 + c^2 + a^2 p^2) \pm \frac{1}{2} \left\{ (a^2 + c^2 + a^2 p^2)^2 - 4a^2 c^2 \right\}^{\frac{1}{2}}.$$

(N.B. This is different from the result obtained by Cyganski & Orr [58].) Once we know γ and K, we can determine α and β from (C.3) and (C.2) respectively.

Problems arise when $\cos \gamma = 0$, in which case (C.2) and (C.3) cannot be used, and c in (C.4) is not defined. To get around this, we can use the fact that

$$\tan \alpha = \frac{b}{d \sin \beta} \quad \text{and} \quad AD - BC = K^2 \cos \alpha \, \cos \beta,$$

and set $c = K \cos \beta$.

Appendix D

Fourier Descriptor Experiment

This appendix describes the two shapes 'c' and 'τ' used in the experiments described in chapters 2 and 4, presents the boundary finding algorithm and reproduces Arbter *et al.*'s [3, 80] equations for obtaining the invariant Fourier descriptors from the vertices of the boundary.

Figure D.1 shows the dimensions of the binary shapes 'c' and 'τ'; they have intensity 1 on a zero intensity background. The figure also shows the numbering scheme used by the boundary tracing algorithm to describe the relative directions of the central pixel's 8-neighbours. The algorithm is described below:

1. Find the starting pixel by scanning each row of the image from left to right, starting at the top row, until a non-zero pixel is found that either has more than two non-zero 8-neighbours or has two 8-neighbours that are 8-neighbours of one another. Set the current pixel equal to the starting pixel.

2. Set *direction* := 0.

3. **If** the pixel in the direction *direction* relative to the current pixel has value 1

   and it is on the edge of the 128x128 image

   > **or** it has more than two 8-neighbours

   > **or** it has only two 8-neighbours, and they are 8-neighbours of one another

   accept it as a boundary point and make it the current pixel.

   else

   direction := (*direction* + 1) modulo 8;

   goto step 3.

4. **If** the current pixel is the starting point

   terminate;

   else

   goto step 3.

Points with exactly two 8-neighbours which are not 8-neighbours of one another are rejected to prevent the algorithm from following thin lines of width one pixel, which in the case of solid images would terminate and cause the algorithm to double back on itself. The vertices of the image are defined as all those pixels at which the boundary changes direction.

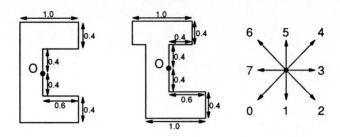

Figure D.1: The dimensions of 'c' and 'τ' used in the experiments, and the direction labels for a pixel's 8-neighbours.

If we have N vertices whose x and y coordinates are given by u_i, v_i, $i = 0, 1, \ldots, N-1$ respectively, then the zero order invariants $I_k = |Q_k|$, $k = 1, 2, \ldots$, are obtained as follows [3] (setting $(u_N, v_N) = (u_0, v_0)$): define the translation normalized coordinates u_i', v_i' as

$$\begin{bmatrix} u_i' \\ v_i' \end{bmatrix} = \begin{bmatrix} u_i - u_{AC} \\ v_i - v_{AC} \end{bmatrix}, \quad \text{where} \quad \begin{bmatrix} u_{AC} \\ v_{AC} \end{bmatrix} = \frac{1}{3A} \sum_{i=0}^{N-1} a_i \begin{bmatrix} u_i + u_{i+1} \\ v_i + v_{i+1} \end{bmatrix}$$

$$\text{and} \quad a_i = u_i v_{i+1} - u_{i+1} v_i, \quad A = \sum_{i=0}^{N-1} a_i.$$

The *signed* area parameterization for polygons is given by

$$t_0 = 0; \quad t_{i+1} = t_i + \frac{1}{2}(u_i' v_{i+1}' - u_{i+1}' v_i'), \quad i = 0, \ldots, N-1.$$

Setting $T = t_N$, $\Phi_{k,i} = e^{-j2\pi k t_i/T}$, $j^2 \equiv -1$, allows one to write the coefficients U_k, V_k of the Fourier series describing the parameterized boundary as

$$\begin{bmatrix} U_k \\ V_k \end{bmatrix} = \frac{1}{2\pi k} \sum_{i=0}^{N-1} \gamma_i \begin{bmatrix} u_{i+1} - u_i \\ v_{i+1} - v_i \end{bmatrix}, \quad \text{where} \quad \gamma_i = \begin{cases} \dfrac{T(\Phi_{k,i+1} - \Phi_{k,i})}{2\pi k(t_{i+1} - t_i)} & \text{if } t_{i+1} \neq t_i; \\[4mm] j\Phi_{k,i} & \text{if } t_{i+1} = t_i. \end{cases}$$

Since the u_i and v_i are real, one need only consider values of $k \geq 1$ (V_0 and U_0 cannot be used to obtain invariants). The zero order absolute invariants $I_k = |Q_k|$, $k = 1, 2, \ldots$ used by Arbter *et al.* [3] were used in our experiments; the Q_k are defined by

$$Q_k = \frac{\Delta_k}{\Delta_p} = \frac{U_k V_p^* - V_k U_p^*}{U_p V_p^* - V_p U_p^*},$$

with V^* denoting the complex conjugate of V. As discussed by Arbter *et al.*, in the presence of noise p should be chosen to make Δ_p as large as possible; with the letters 'c' and 'τ' it was largest when $p = 1$, so this is the value we used. The experiments tested the invariance of the I_k for $k = 2, 3, \ldots, 21$.

Bibliography

[1] Burns J.B., Weiss R.S., and Riseman E.M. The non-existence of general-case view-invariants. In Mundy J.L. and Zisserman A., editors, *Geometric Invariance in Computer Vision*, pages 120–131. MIT Press, 1992.

[2] Dudani S.A., Breeding K.J., and McGhee R.B. Aircraft identification by moment invariants. *IEEE Transactions on Computers*, C-26:39–46, January 1977.

[3] Arbter K., Snyder W.E., Burkhardt H., and Hirzinger G. Application of affine-invariant Fourier descriptors to recognition of 3-D objects. *IEEE Trans. Pattern Anal. & Machine Intell.*, 12:640–647, July 1990.

[4] Reeves A.P., Prokop R.J., Andrews S.E., and Kuhl P.K. Three-dimensional shape analysis using moments and Fourier descriptors. *IEEE Trans. Pattern Anal. & Machine Intell.*, 10(6):937–943, November 1988.

[5] Chen Z. and Ho S.-Y. Computer vision for robust 3D aircraft recognition with fast library search. *Pattern Recognition*, 24(5):375–390, 1991.

[6] Weiss I. Projective invariants of shapes. Technical report, Center for Automation Research, University of Maryland, College Park, MD 20742, U.S.A., January 1988. Also in *Proc. DARPA Image Understanding Workshop*, 1988.

[7] Barrett E.B., Payton P.M., Haag N.N., and Brill M.H. General methods for determining projective invariants in imagery. *CVGIP: Image Understanding*, 53(1):46–65, January 1991.

[8] Barrett E.B., Payton P.M., and Brill M.H. Contributions to the theory of projective invariants for curves in two and three dimensions. *DARPA-ESPRIT Workshop on Applications of Invariance in Computer Vision*, pages 387–424, March 1991.

[9] Forsyth D., Mundy J.L., Zisserman A., Coelho C., Heller A., and Rothwell C. Invariant descriptors for 3-D object recognition and pose. *IEEE Trans. Pattern Anal. & Machine Intell.*, 13(10):971–991, October 1991.

[10] Mundy J.L. and Zisserman A. *Geometric Invariance in Computer Vision*. MIT Press, 1992.

[11] Thompson D.W. and Mundy J.L. Three-dimensional model matching from and unconstrained viewpoint. *Proc. ICRA*, pages 208–220, April 1987.

[12] Wechsler H. Invariance in pattern recognition. In *Advances in Electronics and Electron Physics*, pages 261–322. Academic Press, 1987.

[13] Kanatani K. *Group-Theoretical Methods in Image Understanding*. Springer, Heidelberg, 1990.

[14] Abu-Mostafa Y.S. and Psaltis D. Image normalization by complex moments. *IEEE Trans. Pattern Anal. & Machine Intell.*, PAMI-7:46–55, January 1985.

[15] Teh C-H. and Chin R.T. On image analysis by the method of moments. *IEEE Transactions on Pattern Anal. & Machine Intell.*, 10:496–513, July 1988.

[16] Khotanzad A. and Lu J-H. Classification of invariant image representation using a neural network. *IEEE Trans. Acoustics, Speech & Signal Proc.*, 38:1028–1038, June 1990.

[17] Fenske A. and Burkhardt H. Affine invariant recognition of gray scale objects by Fourier descriptors. *Proceedings of the SPIE*, 1567:53–64, 1991.

[18] Glauser T. and Bunke H. Edge length ratios: an affine invariant shape representation for recognition with occlusions. *11th IAPR Int. Conf. Pattern Recognition, Den Haag, Netherlands*, 1:437–440, August 1992.

[19] Lamdan Y., Schwartz J.T., and Wolfson H.J. Affine invariant model-based object recognition. *IEEE Trans. Robotics & Automation*, 6(5):578–589, October 1990.

[20] Rothwell C.A., Zisserman A., Forsyth D.A., and Mundy J.L. Using projective invariants for constant time library indexing in model based vision. In *British Machine Vision Conference*, pages 62–70. Springer Verlag, Heidelberg, September 1991.

[21] Lamdan Y., Schwartz J.T., and Wolfson H.J. Object recognition by affine invariant matching. *IEEE Int. Conf. Computer Vision and Pattern Rec.*, pages 335–344, June 1988.

[22] Bruckstein A.M. and Netravali A.N. On differential invariants of planar curves and the recognition of partially occluded planar shapes. Technical report, AT&T, July 1990. Also in International Workshop on Visual Form, Capri, May 1991.

[23] Giannakis G.B. and Tsatsanis M.K. Signal detection and classification using matched filtering and higher order statistics. *IEEE Trans. Acoustics, Speech & Signal Proc.*, 38:1284–1296, July 1990.

[24] Minsky M.L. and Papert S.A. *Perceptrons, 2nd ed.* MIT Press, 1988.

[25] Hu M.-K. Pattern recognition by moment invariants. *Proceedings of the IRE*, 49:1428, September 1961.

[26] Hu M.-K. Visual pattern recognition by moment invariants. *IRE Transactions on Information Theory*, IT-8:179–187, February 1962.

[27] Reiss T.H. The revised fundamental theorem of moment invariants. *IEEE Trans. Pattern Anal. Machine Intell.*, 13:830–834, August 1991.

[28] Casasent D. and Psaltis D. Position, rotation, and scale invariant optical correlation. *Applied Optics*, 15:1975–1800, July 1976.

[29] Wechsler H. *Computational Vision*. Academic Press, London, 1990.

[30] Schwartz E.L. Spatial mapping in the primate sensory projection: analytic structure and relevance to perception. *Biological Cybernetics*, 25:181–194, 1977.

[31] Schwartz E.L. Computational anatomy and functional architecture of striate cortex: a spatial mapping approach to perceptual coding. *Vision Research*, 20:645–669, 1980.

[32] Tootell R.B.H., Silverman M.S., Switkes E., and de Valois R.L. Deoxyglucose analysis of retinotopic organization in primate striate cortex. *Science*, 218:902–904, 1982.

[33] Resnikoff H.L. *The Illusion of Reality*. Springer Verlag, 1989.

[34] Lenz R. *Group Theoretical Methods in Image Processing*, volume 413 of *Lecture Notes in Computer Science*. Springer, 1990.

[35] Davis P.J. Plane regions determined by complex moments. *Journal of Approximation Theory*, 19:148–153, 1977.

[36] Abu-Mostafa Y.S. and Psaltis D. Recognitive aspects of moment invariants. *IEEE Trans. Pattern Anal. & Machine Intell.*, PAMI-6:698–706, November 1984.

[37] Teague M.R. Image analysis via the general theory of moments. *J. Opt. Soc. Am.*, 70, August 1980.

[38] Bhatia A.B. and Wolf E. On the circle polynomials of Zernike and related orthogonal sets. *Proc. Cambridge Philosophical Society*, 50:40–48, 1954.

[39] Khotanzad A. and Hong Y.H. Invariant image recognition by Zernike moments. *IEEE Trans. Pattern Anal. & Machine Intell.*, 12:489–497, May 1990.

[40] Maitra S. Moment invariants. *Proceedings of the IEEE*, 67:697–699, April 1979.

[41] Sheng Y. and Arsenault H.H. Experiments on pattern recognition using invariant Fourier-Mellin descriptors. *J. Opt. Soc. Am. A*, 3:771–776, June 1986.

[42] Sheng Y. and Duvernoy J. Circular-Fourier-radial-Mellin transform descriptors for pattern recognition. *J. Opt. Soc. Am.*, 3:885–888, June 1986.

[43] Fuchs A. and Haken H. Pattern recognition and associative memory as dynamical processes in a synergetic system I. *Biological Cybernetics*, 60:17–22, 1988.

[44] Fuchs A. and Haken H. Pattern recognition and associative memory as dynamical processes in a synergetic system II. *Biological Cybernetics*, 60:107–109, 1988.

[45] Fuchs A. and Haken H. Computer simulations of pattern recognition as a dynamical process. In *Neural and Synergetic Computers*. Springer Verlag, 1988.

[46] Fuchs A. and Haken H. Erratum. Pattern recognition and associative memory as dynamical processes in a synergetic system. *Biological Cybernetics*, 60:476, 1989.

[47] Udagawa K., Toriwaki J., and Sugino K. Normalization and recognition of two-dimensional patterns with linear distortion by moments. *Electron. Commun. Japan*, 47(6):34–46, 1964.

[48] Tsai W-H. and Chou S-L. Detection of generalized principal axes in rotationally symmetric shapes. *Pattern Recognition*, 24(2):95–104, February 1991.

[49] Barnard E. and Casasent D. Invariance and neural nets. *IEEE Trans. Neural Networks*, 2(5):498–508, September 1991.

[50] Mardia K.V. and Hainsworth T.J. Statistical aspects of moment invariants in image analysis. *Journal of Applied Statistics*, 16(3):423–435, 1989.

[51] Jiang X.Y. and Bunke H. Simple and fast computation of moments. *Pattern Recognition*, 24(8):801–806, 1991.

[52] Perantonis S.J. and Lisboa P.J.G. Translation, rotation, and scale invariant pattern recognition by high-order neural networks and moment classifiers. *IEEE Trans. Neural Networks*, 3(2):241–251, March 1992.

[53] Giles C.L. and Maxwell T. Learning, invariance, and generalization in high-order networks. *Applied Optics*, 26:4972–4978, December 1987.

[54] Chen K. Efficient parallel algorithms for the computation of two-dimensional image moments. *Pattern Recognition*, 23:109–119, 1990.

[55] Hatamian M. A real-time two-dimensional moment generating algorithm and its single chip implementation. *IEEE Trans. Acoustics, Speech & Signal Proc.*, Acoustics, Speech & Signal Proc.-34:546–553, June 1986.

[56] Zakaria M.F., Vroomen L.J., Zsombor-Murray P.J.A., and van Kessel J.M.H.M. Fast algorithm for the computation of moment invariants. *Pattern Recognition*, 20:639–643, 1987.

[57] Pan Y. A note on efficient parallel algorithms for the computation of two-dimensional image moments. *Pattern Recognition*, 24(9):917, 1991.

[58] Cyganski D. and Orr J.A. Applications of tensor theory to object recognition and orientation determination. *IEEE Trans. Pattern Anal. & Machine Intell.*, PAMI-7:662–673, November 1985.

[59] Abhyankar S.S. Invariant theory and enumerative combinatorics of young tableaux. *DARPA-ESPRIT Workshop on Applications of Invariance in Computer Vision*, pages 55–108, March 1991.

[60] Salmon G. *Higher Plane Curves*. Hodges, Foster and Figgis, Dublin, 1879.

[61] Salmon G. *Lessons introductory to the Modern Higher Algebra, 4th ed.* Hodges, Figgis & Co., Dublin, 1885.

[62] Hilbert D. Über die Theorie der algebraischen Formen. *Math. Ann.*, 36:473–534, 1890.

[63] Hilbert D. Über die vollen Invariantensysteme. *Math. Ann.*, 42:313–373, 1893.

[64] Gurevich G.B. *Foundations of the theory of algebraic invariants.* P. Noordhoff Ltd., Groningen, The Netherlands, 1964.

[65] Dickson L.E. *Algebraic Invariants.* John Wiley & Sons, 1914.

[66] Parshall K.H. Toward a history of nineteenth-century invariant theory. In *The History of Modern Mathematics*, pages 157–208. Academic Press, 1989.

[67] Reiss T.H. and Rayner P.J.W. On generating features invariant to linear transformations in two and three dimensions. Technical Report CUED/F-INFENG/TR.87, Cambridge University, Cambridge, England, November 1991.

[68] Reiss T.H. Features invariant to linear transformations in 2D and 3D using tensors. *11th IAPR Int. Conf. on Pattern Recognition, The Hague, Netherlands*, 3:493–496, August 1992.

[69] Clebsch A. *Theorie der binären algebraischen Formen.* B.G. Teubner, Leipzig, 1872.

[70] Rayner P.J.W. and Lynch M.R. A new connectionist model based on a non-linear adaptive filter. *Proceedings IEEE Int. Conf. Acoustics, Speech & Signal Proc.*, 2:1191–1194, 1989.

[71] Cayley A. A second memoir upon quantics. *Philosophical Transactions of the Royal Society of London*, 146:101–126, 1856.

[72] Sylvester J.J. Proof of the hitherto undemonstrated fundamental theorem of invariants. *Philosophical Magazine*, 5:178–188, 1878.

[73] Quan L., Gros P., and Mohr R. Invariants of a pair of conics revisited. In *British Machine Vision Conference*, pages 71–77. Springer Verlag, Heidelberg, September 1991.

[74] Weiss I., Meer P., and Dunn S.M. Robustness of algebraic invariants. *DARPA-ESPRIT Workshop on Applications of Invariance in Computer Vision*, pages 345–357, March 1991.

[75] Dirilten H. and Newman T.G. Pattern matching under affine transformations. *IEEE Trans. Comput.*, C-26:314–317, March 1977.

[76] Bamieh B. and de Figueiredo R.J.P. A general moment-invariants/attributed-graph method for three-dimensional object recognition from a single image. *IEEE J. Robotics & Automation*, RA-2(1):31–41, March 1986.

[77] Taubin G. and Cooper D.B. Object recognition based on moment (or algebraic) invariants. In Mundy J.L. and Zisserman A., editors, *Geometric Invariance in Computer Vision*, pages 375–397. MIT Press, 1992.

[78] Cyganski D. and Orr J.A. Object recognition and orientation determination by tensor methods. In T.S. Huang, editor, *Advances in Computer Vision and Image Processing*, pages 101–144. JAI Press Inc. ISBN: 0-89232-636-2, 1988.

[79] Leu J.-G. Shape normalization through compacting. *Pattern Recognition Letters*, 10:243–250, October 1989.

[80] Arbter K. *Affinvariante Fourierdeskriptoren ebener Kurven*. PhD thesis, Technische Universität Hamburg-Harburg, Germany, 1990.

[81] Burkhardt H., Fenske A., and Schulz-Mirbach H. Invariants for the recognition of planar contour and gray-scale images. In Oxford University H. Burkhardt, Technische Universität Hamburg; A. Zisserman, editor, *Invariants for Recognition, ESPRIT Basic Research Workshop, ECCV '92, Italy*, pages 1–26. May 1992.

[82] Jin Q. and Yan P. A new method of extracting invariants under affine-transform. *IAPR Conference on Pattern Recognition, Den Haag, Netherlands*, 1:742–745, August 1992.

[83] Van Gool L., Wagemans J., Vandeneede J., and Oosterlinck A. Similarity extraction and modelling. In *3rd IEEE Int. Conf. Computer Vision, Osaka, Japan*, pages 530–534, December 1990.

[84] Brill M.H., Barrett E.B., and Payton P.M. Projective invariants for curves in two and three dimensions. In Mundy J.L. and Zisserman A., editors, *Geometric Invariance in Computer Vision*, pages 193–214. MIT Press, 1992.

[85] Barrett E.B., Brill M.H., Haag N.N., and Payton P.M. Invariant linear methods in photogrammetry and model-matching. In Mundy J.L. and Zisserman A., editors, *Geometric Invariance in Computer Vision*, pages 277–292. MIT Press, 1992.

[86] Costa M., Haralick R., Phillips T., and Shapiro L. Optimal affine-invariant point matching. *SPIE Applications of Artificial Intelligence VII*, 1095:515–530, 1989.

[87] Hopcroft J.E., Huttenlocher D.P., and Wayner P.C. Affine invariants for model-based recongnition. In Mundy J.L. and Zisserman A., editors, *Geometric Invariance in Computer Vision*, pages 354–374. MIT Press, 1992.

[88] Hummel R. and Wolfson H. Affine invariant matching. *DARPA Image Understanding Workshop*, April 1988.

[89] Reiss T.H. and Rayner P.J.W. Object recognition using algebraic and differential invariants. Technical Report CUED/F-INFENG/TR.97, Cambridge University Engineering Department, Cambridge, England, May 1992.

[90] Matthews V.J. Adaptive polynomial filters. *IEEE Signal Processing Magazine*, 8:10–26, July 1991.

[91] Specht D.F. Generation of polynomial discriminant functions for pattern recognition. *IEEE Transactions on Electronic Computers*, C-16:308–319, June 1967.

[92] Specht D.F. Probabilistic neural networks and the polynomial adaline as complementary techniques for classification. *IEEE Trans. Neural Networks*, 1:111–121, March 1990.

[93] Ivakhnenko A.G. The group method of data handling — a rival of the method of stochastic approximation. *Soviet Automatic Control*, 3, 1968.

[94] Farlow S. *Self-Organizing Methods in Modeling*. Marcel Dekker, NY, 1984.

[95] Molnar D.O. Modeling accuracy improvements for the group method of data handling. *1st IEEE Int. Conference on Neural Networks*, 4:839–846, 1987.

[96] Nilsson N.J. *Learning Machines*. McGraw Hill, 1965.

[97] Duda R.O. and Hart P.E. *Pattern Classification and Scene Analysis*. John Wiley & Sons, 1973.

[98] Rayner P.J. and Lynch M.R. Complexity reduction in Volterra connectionist modelling by consideration of output mapping. *Proceedings IEEE Int. Conf. Acoustics, Speech & Signal Proc.*, 2:885–888, 1990.

[99] Reiss T.H. Learning image invariants with the Volterra Connectionist Model. *IEEE Int. Conf. Acoustics, Speech & Signal Proc.*, 2:1045–1048, May 1991.

[100] Lynch M.R., Rayner P.J., and Holden S.B. Removal of degeneracy in adaptive Volterra networks by dynamic structuring. *IEEE Int. Conf. Acoustics, Speech & Signal Proc.*, 1991.

[101] Lynch M.R., Holden S.B., and Rayner P.J.W. Complexity reduction in Volterra connectionist networks using a self-structuring LMS algorithm. *IEE 2nd International Conf. on 'Artificial Neural Networks', Bournemouth, England*, November 1991.

[102] Geva S. and Sitte J. Adaptive nearest neighbor classification. *IEEE Trans. Neural Networks*, 2:318–322, March 1991.

[103] Van Gool L.J., Moons T., Pauwels E., and Oosterlinck A. Semi-differential invariants. In Mundy J.L. and Zisserman A., editors, *Geometric Invariance in Computer Vision*, pages 157–192. MIT Press, 1992.

[104] Kanatani K. Camera rotation invariance of image characteristics. *Computer Vision, Graphics, and Image Processing*, 39:328–354, 1987.

[105] Mohr R., Sparr G., and Faugeras O. Multiple image invariants. In Oxford University H. Burkhardt, Technische Universität Hamburg; A. Zisserman, editor, *Invariants for Recognition, ESPRIT Basic Research Workshop, ECCV '92, Italy*, pages 109–132. May 1992.

[106] Van Gool L.J., Brill M.H., Barrett E.B., Moons T., and Pauwels E. Semi-differential invariants for nonplanar curves. In Mundy J.L. and Zisserman A., editors, *Geometric Invariance in Computer Vision*, pages 293–309. MIT Press, 1992.

[107] Brown C. Numerical evaluation of differential and semi-differential invariants. In Mundy J.L. and Zisserman A., editors, *Geometric Invariance in Computer Vision*, pages 215–227. MIT Press, 1992.

[108] Weiss I. Differential invariants without derivatives. *11th IAPR Int. Conf. on Pattern Recognition*, 3:394–398, August 1992.

[109] Coelho C., Heller A., Mundy J.L., Forsyth D.A., and Zisserman A. An experimental evaluation of projective invariants. In Mundy J.L. and Zisserman A., editors, *Geometric Invariance in Computer Vision*, pages 87–104. MIT Press, 1992.

[110] Sanfeliu A., Llorens A., and Emde W. Sensibility, relative error and error probability of projective invariants of planar surfaces of 3D objects. *11th IAPR Int. Conf. on Pattern Recognition, Den Haag, Netherlands*, 1:328–331, August 1992.

[111] Carlsson S. Projectively invariant decomposition of shapes. In Mundy J.L. and Zisserman A., editors, *Geometric Invariance in Computer Vision*, pages 267–273. MIT Press, 1992.

[112] Forsyth D., Mundy J.L., Zisserman A., and Brown C.M. Projectively invariant representations using implicit algebraic curves. In *1st European Conf. Computer Vision*, pages 427–436. Springer Verlag, Heidelberg, 1990.

[113] Semple J.G. and Kneebone G.T. *Algebraic Projective Geometry*. Oxford University Press, 1952.

[114] Bookstein F.L. Fitting conic sections to scattered data. *Computer Graphics and Image Proc.*, 9:56–71, 1979.

[115] Paton K.A. Conic sections in chromosome analysis. *Pattern Recognition*, 2:39, 1970.

[116] Paton K.A. Conic sections in automatic chromosome analysis. *Machine Intelligence*, 5:411, 1970.

[117] Biggerstaff R.H. Three variations in dental arch form estimated by a quadratic equation. *J. Dent. Res.*, 51:1509, 1972.

[118] Albano A. Representation of digitized contours in terms of conic arcs and straight-line segments. *Computer Graphics and Image Proc.*, 3:23, 1974.

[119] Cooper D.R. and Yalabik N. On the computational cost of approximating and recognizing noise-perturbed straight lines and quadric arcs in the plane. *IEEE Trans. Computers*, C-25:1020, 1976.

[120] Kapur D. and Mundy J.L. Fitting affine invariant conics to curves. In Mundy J.L. and Zisserman A., editors, *Geometric Invariance in Computer Vision*, pages 252–266. MIT Press, 1992.

[121] Reiss T.H. Object recognition using algebraic and differential invariants. *Signal Processing, to appear*, 32(3), September 1993.

[122] Zisserman A. and Rothwell C. Personal communication, 1992.

[123] Lowe D.G. *Perceptual Organization and Visual Recognition*. Kluwer, Boston, 1985.

[124] Lowe D.G. Fitting parametrized three-dimensional models to images. *IEEE Trans. Pattern Anal. Machine Intell.*, 13:441–450, May 1991.

[125] Dhome M., Richetin M., Lapresté J.-T., and Rives G. Determination of the attitude of 3-D objects from a single perspective view. *IEEE Trans. Pattern Anal. Machine Intell.*, pages 1265–1278, December 1989.

[126] Grimson W.E.L. The combinatorics of heuristic search termination for object recognition in cluttered environments. *IEEE Trans. Pattern Anal. & Machine Intell.*, 13(9):920–935, September 1991.

[127] Grimson W.E.L. and Huttenlocher D.P. On the verification of hypothesized matches in model-based recognition. *IEEE Transactions on Pattern Anal. & Machine Intell.*, 13(12):1201–1213, December 1991.

[128] Wolfson H.J. and Lamdan Y. Transformation invariant indexing. In Mundy J.L. and Zisserman A., editors, *Geometric Invariance in Computer Vision*, pages 335–353. MIT Press, 1992.

[129] Gavrila D.M. and Groen F.C.A. 3D object recognition from 2D images using geometric hashing. *Pattern Recognition Letters*, 13:263–278, April 1992.

[130] Huttenlocher D.P. Fast affine point matching: An output-sensitive method. *IEEE Intl. Conf. Computer Vision & Pattern Rec., Hawaii.*, pages 263–268, June 1991.

[131] Van Gool L., Moons T., Pauwels E., and Oosterlinck A. Semi-differential invariants. *DARPA-ESPRIT Workshop on Applications of Invariance in Computer Vision*, pages 359–386, March 1991.

[132] Jacobs D.W. Optimal matching of planar models in 3D scenes. *IEEE Intl. Conf. Computer Vision & Pattern Rec., Hawaii.*, pages 269–274, June 1991.

[133] Sylvester J.J. Memoir on the dialytic method of elimination. *Philosophical Magazine*, 21:534–539, 1842.

[134] Cayley A. On linear transformations. *Cambridge and Dublin Mathematical Journal*, 1:104–122, 1846.

[135] Sylvester J.J. On a rule for abbreviating the calculation of the number of in- or co-variants of a given order and weight in the coefficients of a binary quantic of a given degree. *Messenger of Mathematics*, 8:1–8, 1879.

[136] Sylvester J.J. Tables of the generating functions and groundforms for the binary quantics of the first ten orders. *American Journal of Mathematics*, 2:223–251, 1879.

[137] Sylvester J.J. Tables of the generating functions and groundforms for simultaneous binary quantics of the first four orders taken two and two together. *American Journal of Mathematics*, 2:293–306, 324–329, 1879.

Printing: Weihert-Druck GmbH, Darmstadt
Binding: Buchbinderei Schäffer, Grünstadt